Forensic Psychology 3CC3 Handbook

Dick Day

Winter, 2015-2016

Copyright 2015
Richard B. Day

Table of Contents

Course Outline . . . . Page ii

Introduction to this Handbook . . . . Page v

Introduction to Forensic Psychology . . . . Page 1

Interviews and Interrogation . . . . Page 13

Deception Detection . . . . Page 29

Eyewitness Testimony . . . . Page 47

Criminal Profiling . . . . Page 79

Jury Psychology . . . . Page 105

Assessment . . . . Page 125
    Competence and Criminal Responsibility . . . . Page 125
    Violence Risk . . . . Page 151

Psychopathy . . . . Page 163

Psychology 3CC3: Forensic Psychology
Course Outline, September - December 2015

Time: **Tues., Wed., Fri. 3:30 pm**
Classroom: **JHE-376**
Course web site: **http://intropsych.mcmaster.ca/psych3cc3**
plus the course pages on Avenue 2 Learn

Instructor: **Dick Day**
Office: Psych Bldg Room 404
Phone: 525-9140 ext. 23006
Email:dayrich@mcmaster.ca

Required Text: Pozzulo, Bennell, and Forth. *Forensic Psychology, 4th Edition.* Pearson Education Canada

Recommended Text: Day, R. B. *Forensic Psychology 3CC3 Handbook, 2015-16 Term 2*. McMaster Custom Courseware, 2015.

Course Objectives

The goals of the course are to familiarize students with the nature and scope of forensic psychology, to describe in more detail some of the specific responsibilities and activities of forensic psychologists, and to familiarize students with the theory and empirical evidence that guide forensic psychologists in their work within the legal system.

Course Topics

The table below shows the topics we will cover in this course, and the order in which we will discuss them. The specific weeks in which we will cover these topics may vary from the dates given below, depending on how deep and active our discussion of each topic becomes. So take the dates below as approximate and subject to revision.

Week of:	Topic	Text Readings
Jan. 4	Introduction to Forensic Psychology	Chapt. 1
Jan. 11 Jan. 18 Jan. 25	Interview and Interrogation Procedures; Detecting Deception; Eyewitness Testimony	Chapt. 3 (pp. 58-77) Chapt. 4 (pp.92-112); Chapt. 5; Chapt. 6 (pp. 154-173)
Feb. 1 * Feb. 8	Profiling in Criminal Psychology	Chapt. 3 (pp. 78-90)
Feb. 15	Break Week	
Feb. 22 Feb. 29 Mar. 7	The Psychology of Juries	Chapt. 7
* Mar. 14 Mar. 21 Mar. 28	Assessment of Competence, Criminal Responsibility, Risk of Violence	Chapt. 8; Chapter 10
Apr. 4	Psychopathy **(Classes end Friday, April 8th)**	Chapt. 11
Exam Period	Final Examination	

Course Evaluation: There will be two non-cumulative in-class midterm tests, worth 30% each, and a cumulative final exam worth 40%.

Each in-class test will consist of approximately 40 multiple-choice questions. You will have the full 50 minutes of the class period to complete each test.

The final exam will consist of approximately 80 multiple-choice questions and will be cover all the material in the course. You will have two hours for this test, which will take place during the December examination period. The **tentative dates** for the in-class tests are as follows (and are marked on the course outline with an asterisk next to the week in which they will take place):

> **In-class Test #1**: Tuesday, February 9th
> **In-class Test #2**: Tuesday, March 15th

The **actual** dates of the in-class tests will depend on the availability of additional testing rooms (if required), and will be announced in class - and posted on the course website - **at least 10 days** prior to the actual test date. **IT IS YOUR RESPONSIBILITY TO MAKE SURE THAT YOU GET THIS INFORMATION.**

Missed Work: If you miss one of the in-class tests for documented medical or compassionate reasons, you should complete a Missed Work form in the office of the Associate Dean (Studies) of your Faculty. Once your Associate Dean has accepted your reason for absence, the remaining in-class test and the final exam will be reweighted (40%, 60%, respectively) to cover the missed in-class test.

Final Grade Calculation and Adjustment: The final mark in Psychology 3CC3 will be computed by applying the following formula to the percentage scores on Test1, Test2, and Final Exam:

$$(Test1\% \times .30) - (Test2\% \times .30) - (Exam\% \times .40) = \text{Final Course Mark}$$

Apart from excused absences from an in-class test, every student will be assessed using the weighting formula shown above - with one exception:

In assigning final letter grades for the course I look at the pattern of performance over the two in-class tests and the final exam. If the overall average, as calculated by the formula above, is on the borderline of the next higher letter grade (e.g. 49%, 66%, or 84%) and if the marks on both the final exam and one in-class midterm test are at the next higher level (e.g., D-, C+, A), then I will assign the next higher letter grade.

Apart from this one final adjustment, final course grades in Psych 3CC3 will not be changed unless they have been calculated incorrectly. I do not respond to phone, email or in-person requests to increase correctly-calculated grades in this course.

Note that although midterm and exam marks, and final course averages are posted on Avenue as well as on the intropsych Grades Lookup, **ONLY THE NUMERICAL POSTINGS ON THE GRADES LOOKUP REFLECT THE CORRECT CALCULATION OF YOUR MARKS AND FINAL AVERAGE.** The letter grades posted on Avenue and the Grades Lookup will be identical, and equally correct.

The instructor and university reserve the right to modify elements of the course during the term. The university may change the dates and deadlines for any or all courses in

extreme circumstances. **If either type of modification becomes necessary, reasonable notice and communication with the students will be given with explanation and the opportunity to comment on changes. It is the responsibility of the student to check their McMaster email and course websites weekly during the term and to note any changes**.

Academic Integrity and Academic Dishonesty: Academic dishonesty consists of misrepresentation by deception or by other fraudulent means and can result in serious consequences, e.g. the grade of zero on an assignment, loss of credit with a notation on the transcript (notation reads: "Grade of F assigned for academic dishonesty"), and/or suspension or expulsion from the university.

It is your responsibility to understand what constitutes academic dishonesty. For information on the various kinds of a academic dishonesty please refer to the Academic Integrity Policy, specifically Appendix 3, located at

http://www.mcmaster.ca/senate/academic/ac_integrity.htm

The following illustrates only three forms of academic dishonesty:
1. **Copying or using unauthorized aids on tests and examinations**.
2. Plagiarism, e.g. the submission of work that is not one's own or for which other credit has been obtained.
3. Improper collaboration in group work.

Grading in Psychology 3CC3

The final mark in Psychology 3CC3 will be computed by applying the following formula to the percentage scores on Midterm 1, Midterm 2, and the Final Exam:

(Midterm1% x .30) + (Midterm2% x .30) + (Final Exam% x .40) = Final Course Percentage

The Final Course Percentage will be translated into a letter grade according to the following table of equivalence:

% Score	Letter	% Score	Letter	% Score	Letter
90 - 100	A+	70 - 72	B-	53 - 56	D
85 - 89	A	67 - 69	C+	50 - 52	D-
80 - 84	A-	63 - 66	C	0 - 49	F
77 - 79	B+	60 - 62	C-		
73 - 76	B	57 - 59	D+		

Using This Handbook and Other Course Materials

This Handbook

This Handbook contains the outline notes that I work from during lectures - at least the version of those notes that was current when the Handbook was printed. I provide a Handbook for my courses for several reasons.

First, I want to make sure that you do not have to be constantly scribbling (or typing) in notes during the lecture, but can spend most of your time thinking about the material we are covering. It makes little sense to me to have you taking dictation from material that I already have in digital form, and the best way to master the material in any course is to stay engaged with it for as long as possible.

Second, I want to make sure that you don't miss anything, either because you were unable to attend a particular lecture, or because you forgot or were unable to get everything down as I was rattling on.

Just a few more words about this Handbook. Keep in mind that **THIS HANDBOOK IS NOT THE OFFICIAL CONTENT OF THE COURSE.** The official course content is whatever we talk about in lecture. You will often find that there is material in this Handbook that is not mentioned in lecture. Any material in the Handbook that is not mentioned in lecture **IS NOT PART OF THE COURSE CONTENT THIS TERM.**

On the other hand, new information relevant to Forensic Psychology is constantly being published, and now and again I will mention in lecture a new study or theory that does **NOT** appear anywhere in this Handbook. Any such material **IS** part of the official course content, and may appear on midterm tests and/or the final exam.

Course Website

Copies of the course outline in PDF format, together with a number of other course materials, can be found on the course website: http://intropsych.mcmaster.ca/psych3cc3, (abbreviated below as "intro3cc3") **and can also be found on McMaster's learning management system, Avenue to Learn** (abbreviated below as "A2L"). Information about the contents of the course website(s) can be found below.

Course Outline

There are links on the course home page (on both course websites) to copies of the course outline in PDF format. I will not be providing paper copies of the outline in class. If changes are made to the course outline, those changes will be announced in class, and will appear shortly thereafter on both versions of the posted course outline.

If changes to the outline are announced, make sure to consult and/or download the outlines posted on the web to make sure you have the updated version. This is especially important for in-class test dates, which will not be known for certain until a few weeks into the course.

3CC3 Forum

On the A2L 3CC3 home page there is a tab or menu item for "Discussions". This is where all important announcements about the course will be posted (on the "Course Info and Instructor Announcements" section), and where you can post any questions or comments about the course on the "Comments and Questions" section. You should check the Forum for relevant postings at least once per week.

Lecture Materials

On the intro3cc3 site, you will find copies of the PowerPoint slides that I use in lecture under the link "Lecture Materials". The slides are posted in both PowerPoint 2003 (.PPT) format, and as a much larger Adobe Acrobat (.PDF) file.

I use the slides both to illustrate material from the lectures, and to remind me of what I want to say next so I won't wander too far off course during lectures. They are **NOT** intended to provide a full and accurate summary of the material in the lecture. There will be a number of important points in lecture that will not appear on the slides, either because these points do not lend themselves to graphic illustration or because I don't want the slides to be overcrowded with text. There may also be slides on which not all the points are of equal importance.

The point I want to stress is that **you should not use the posted PowerPoint slides as a substitute for your notes, or the material in this Handbook, as a way of mastering the course material.** The slides are a general guide to the course material, and usually (but not always) contain a mention or illustration of the most important points. To be sure that you have learned everything you need to know, compare your knowledge with the content of your class notes, or with the content of the lecture recordings.

The "Lecture Materials" page may also contain copies of the midterm tests once they have been written.

The same course materials will be found on the A2L site under "Content".

Grades Lookup

On the intro3cc3 site, the "Grades Lookup" link provides access to a complete list of your marks in the course once you enter your student ID number. It will also show the answers you selected on each of the midterm tests, and allow you to compare your answers with the correct answers for each test.

Note that ANYONE who knows your ID number can get access to your course marks and final grade through this page. If you are concerned that this might happen, and do not want your course marks posted online, email me at any time and let me know. I will arrange to have your marks information removed from the online grades database. If your marks are not posted on intro3cc3, you will still be able to get them from the "Grades" section of A2L site, though you will not be able to compare your midterm answers with the master on A2L.

Online Sample Quiz

This link on the intro3cc3 home page provides access to a sample quiz containing multiple-choice questions similar or identical to those that will appear on the in-class midterm tests and on the final exam. You can select questions specific to any content area(s) of the course, and choose how many questions (3, 5, or 10) will be shown per page. Once you have chosen and submitted your answers, you will be given feedback on whether your answers were correct.

Please note the caution on the main quiz page: "The sample quiz is **NOT** a study guide; it should be used only **AFTER** you have studied the material covered by the quiz questions, as a way of assessing your understanding of the material, and identifying content areas where you need to spend a bit more time. Trying to memorize the answers to all the questions in the hopes that they will show up on the in-class tests or final exam won't help you at all."

Several other cautions about using the online quiz:

As the quiz page indicates, about 1% of this time the quiz messes up and provides wrong or meaningless feedback about the accuracy of your answers. When this happens, you should log out (using the "Log Out" button on top of each quiz page) and start again. **DO NOT RELY ON THE QUIZ TO PROVIDE CORRECT ANSWERS TO THE QUESTIONS. ALWAYS RELY ON YOUR CLASS NOTES OR THIS HANDBOOK** to determine the correct answers to online quiz questions.

Note also that the online quiz does not contain any questions from the assigned readings in the text.

Because the material content of the course changes a bit from term to term, the online quiz may contain questions about material that is **NOT** part of this term's course content. Simply ignore those questions.

3CC3 Links

This page on the intro3cc3 site contains any useful or interesting links to external sites. I haven't updated this page recently, and there isn't much on it at the moment. If you find any sites that you think merit mention, let me know.

Search the Site

This page on the intro3cc3 site allows you to search for any text on the site. Because much of the content of the site is located in PowerPoint or PDF files, this isn't as useful as it might be otherwise, but give it a try if you're looking for something specific.

Contact Me

This link on the intro3cc3 home page takes you to a list of my office hours for the current month, and a link to my office cam so you can see whether I am in or not. I sometimes make changes to my office hours during the month, so if you are planning a visit to my office, check here first to make sure I will be available when you stop by. On the A2L site, office hours for the month appear on the Calendar.

Lecture Recordings

Recordings of all course lectures will be posted on the A2L website, typically within a day or two of their delivery. The recordings are posted in both MP3 and WMA formats, and can be found under "Content".

Our classroom (JHE-376) is also set up for combined audio/visual recording in the Echo format, and a link to those recordings will be posted on A2L under "Content", "Lecture Recordings".

On rare occasions, I forget to bring my recorder to class, or there is a technical glitch that results in a missed recording. Similarly, the Echo A/V recording system occasionally (but rarely) may experience a glitch, so **don't count on there always being a recording available for any particular class!**

Introduction To Forensic Psychology

I. Introduction to the course:

   A. We will describe what forensic psychology is, and what forensic psychologists do. What sort of training and experience do they have, what are their professional activities and responsibilities.

   B. Spend most time describing research from **scientific psychology** that forensic psychologists rely on. Usually, time spent on any topic related to amount of research in area - not always commensurate with amount of time that forensic psychologists spend performing duties in that area.

   C. Forensic psychologists work in both civil and criminal cases, but we will focus on criminal not civil law.
      1. Most relevant psychological research done re criminal trials: Stakes not as high in civil as in criminal trials, where outcome could mean jail term or (in U.S.) execution.
      2. Criminal cases are more interesting to most people

II. Your exposure to forensic psychology

   A. Forensic psychology course different because it is the first (and nearly only) course in department that deals explicitly with application of psychology to real-world issues rather than with scientific study of phenomena that might have direct practical applications.

   B. Usually, exposure to most areas of scientific - and even applied - psychology, comes in psychology courses.

   C. First exposure to forensic psychology probably came from media. For most, through movies.

   D. Just about the only depiction of forensic psychology is offender profiling. Perhaps appropriate, since offender profiling was first forensic application of psychology.
      1. Offender profiling first appeared in fiction in 1842 ← Poe
      2. First appeared in reality in Whitechapel murder case in London (1888). ← Jack the Ripper

   E. High profile of offender profiling is ironic for two reasons. This is ironic since offender profiling::
      1. Is activity in which fewest forensic psychologists are involved.
      2. Is least scientific of all forensic psychology activities

III. What is forensic psychology?

   A. The word **forensic**' is from Latin 'forensis' ('public'), derived from Latin 'forum', central public meeting place in Roman city where political and economic business conducted.
      1. Entered English in the 16th or 17th century, meaning related to the law.
      2. Now generally means application of knowledge in some field to issues re legal system, especially criminal justice system.

   B. Many fields of physical, biological and social science have relationships with the law:
      1. **Anthropology**: Identification of victims from skeletal remains; facial reconstruction. We have strong representation in this field here at Mac.
         a. Determine age, sex, height (from individual bones)
         b. Presence of diseases, nature of environment in which they grew up
      2. Biology:
      3. Chemistry
      4. Engineering
      5. Medicine

IV. How important is forensic evidence in criminal cases?

   A. Fewer than 10% of criminal cases involve forensic evidence

   B. Only about 50% of the forensic evidence collected is actually analyzed

   C. Many forensic techniques are not truly scientific, but require expert judgment that includes a subjective component that leaves room - often considerable room - for error:
      1. **Fingerprint comparisons**:
         a. **Haber & Haber (2008)** after a review of the literature, conclude that the research has not yet been conducted that can validate fingerprint comparison methods.
         b. Error rates for expert matches are unknown: How often are experts wrong?
         c. Errors in the Brandon Mayfield case (**Office of the Inspector General, 2006**)
      2. **Tool marks analysis**:
      3. **Bite marks analysis**:
      4. **Ballistic comparisons**:
         a. **Biasotti (1959)**: Found high error rates for matching bullets to the weapons that fired them. Only about 38% of bullets fired from the same gun could be correctly matched, and for bullets fired from different guns about 20% were judged as matching. They concluded that "even under such ideal conditions [as the

experimental test firings] the average percent match for bullets from the same gun is low and the percent match for bullets from different guns is high..."

5. **Hair and fiber analysis**:
   a. In April 2015, the U.S. Justice Department and the FBI acknowledged that "nearly every examiner in an elite FBI forensic unit gave flawed testimony in almost all trials in which they offered evidence against criminal defendants over more than a two-decade period before 2000. Of 28 examiners with the FBI Laboratory's microscopic hair comparison unit, 26 overstated forensic matches in ways that favored prosecutors in more than 95 percent of the 268 trials reviewed so far, according to the National Association of Criminal Defense Lawyers (NACDL) and the Innocence Project, which are assisting the government with the country's largest post-conviction review of questioned forensic evidence. The cases include those of 32 defendants sentenced to death. Of those, 14 have been executed or died in prison, the groups said under an agreement with the government to release results after the review of the first 200 convictions. The FBI errors alone do not mean there was not other evidence of a convict's guilt. Defendants and federal and state prosecutors in 46 states and the District are being notified to determine whether there are grounds for appeals. Four defendants were previously exonerated." (Washington Post, April 18, 2015:) [www.washingtonpost.com/local/crime/fbi-overstated-forensic-hair-matches-in-nearly-all-criminal-trials-for-decades/2015/04/18/39c8d8c6-e515-11e4-b510-962fcfabc310_story.html]
   b. "In one particularly shocking case from 1978, two FBI-trained hair analysts who helped in the prosecution of a murder case couldn't even tell the difference between human hair and dog hair. The case involved a murder in Washington D.C. that year. The victim, a cab driver, was robbed and killed in front of his home. Before long, police centered upon Santae Tribble, then a 17-year-old local from the neighborhood, as a suspect. Tribble maintained his innocence. But no matter what he said and how much his friends vouched, two FBI forensics experts claimed that a single strand of hair recovered near the scene of the crime matched Tribble's DNA. Thanks to that evidence, which was groundbreaking at the time, Tribble was found guilty and sentenced to 20 years to life in prison after 40 minutes of jury deliberation, reported the Washington Post. He would go on to serve 28 years until the truth came out: an independent analysis found that the FBI testimony was flawed. Not a single hair that was found on the scene matched his DNA. After attorneys brought the evidence to the courts, Tribble was exonerated of the crime, though he'd already been released from prison. "The Court finds by clear and convincing evidence that he did not commit the crimes he was convicted of at trial," a judge wrote in the certificate of innocence released at the time, in 2012. It gets worse. Not only did none of the hairs presented as evidence in trial belonged to Tribble, the private lab found that one of the hairs actually came from a dog. "Such is the true state of hair microscopy," Sandra K. Levick, Tribble's lawyer, wrote at the time, in 2012. "Two FBI-trained analysts. could not even distinguish human hairs from canine hairs." Tribble's case in not unique. In a Washington Post story released over the weekend, officials from the FBI and the Justice Department acknowledged the extent of their flawed use of hair forensics prosecutions prior to

2000. The numbers are staggering. Over 95 percent of the cases involving hair evidence that the FBI has reviewed so far contained flawed testimony-257 out of 268 cases. Thirty-two of these flawed cases involve inmates who are currently on death row. In 14 cases, the inmates have already been executed, or died in prison, The Post's Hsu reports. About 1,200 cases across the country still need to be reviewed. Now, the Post says that "federal and state prosecutors in 46 states and the District are being notified to determine whether there are grounds for appeals." Inside the capitol, the only jurisdiction where investigators have re-investigated all the FBI hair-related convictions prior to 2000, three of seven defendants have been exonerated since 2009. Inmates in another two cases had already been exonerated before the review. Specifically, the "flaws" mean that FBI experts gave "statements exceeding the limits of science," and many instances where the shaky scientific ground that the statements were dependent on were not shared with defense attorneys. To put it simply, the testimony was presented as fact, when it was not. In a joint statement to Hsu, the FBI and the Justice Department said they "are committed to ensuring that affected defendants are notified of past errors and that justice is done in every instance. The Department and the FBI are also committed to ensuring the accuracy of future hair analysis testimony, as well as the application of all disciplines of forensic science." (Fusion.net April 21, 2015: http://fusion.net/story/123382/fbi-forensics-once-brought-dog-hair-to-a-mans-murder-trial-to-use-against-him/]

V. General comment on forensic evidence. (**O'Brien, 2010**)

    A. Prosecutors talk about "the CSI effect": Jurors in criminal trials expect prosecutors to prsent forensic evidence, since TV shows like those in the CSI series always hinge on forensic evidence - as real trials seldom do.

    B. A 2006 survey of 1,000 criminal jurors in Michigan found that 46% expected some sort of forensic evidence in every criminal trial: 75% expected such evidence in murder trials. (**Shelton, Barak & Kim, 2007**)
       1. 22% expected DNA evidence
       2. 36% expected fingerprint evidence
       3. 32% expected ballistic or firearms evidence
       4. "In general, frequent CSI watchers had higher expectations for all kinds of evidence than did non-CSI watchers. This means that in all categories of evidence, both scientific and non-scientific, the evidentiary expectations of CSI watchers were consistently higher than those of non-CSI watchers. CSI watchers also had higher expectations about scientific evidence that is more likely to be relevant to a particular crime than non-CSI watchers, and they had lower expectations about evidence that is less likely to be relevant to a particular crime than did non-CSI watchers." (P. 363)

    C. A 2010 study found that potential jurors trust forensic evidence strongly, considering it more reliable than to be far more reliable than the testimony of police officers, eyewitnesses, or crime victims.

D. **Baskin, I. & Sommers, D. (2010)** examined influence of forensic evidence on 400 homicide case outcomes.
1. Only 13.5% of cases had forensic evidence linking suspect to crime scene or victim. Conviction in these cases slightly higher than for all other cases.
2. Biological evidence in 38% of cases; fingerprints in 28%; DNA in 4.5% of homicides.
3. None of the forensic evidence variables significantly influenced criminal justice outcomes, suggesting that forensic evidence is does not determine outcome in homicide cases.

VI. History of Forensic Psychology

A. 1842: "**Edgar Allan Poe** publishes "The Murders in the Rue Morgue", first fictional detective story. Starts interplay between genuine development of forensic science and development of the fictional detective/criminalist.

B. 1887: "**Arthur Conan Doyle** publishes first Sherlock Holmes story ("A Study in Scarlet") in Beeton's Christmas Annual of London."

C. 1888: Dr. Thomas Bond provides the first offender profile in London's Whitechapel murders case: ""*The murderer must have been a man of great physical strength and of great coolness and daring. He must in my opinion be a man subject to periodical attacks of Homicidal and Erotic mania. The character of the mutilations indicate that the man may be in a condition sexually, that may be called Satyriasis. It is of course possible that the Homicidal impulse may have developed from a revengeful or brooding condition of the mind, or that religious mania may have been the original disease but I do not think either hypothesis is likely. The murderer in external appearance is quite likely to be a quiet inoffensive looking man probably middle-aged and neatly and respectably dressed. I think he must be in the habit of wearing a cloak or overcoat or he could hardly have escaped notice in the streets if the blood on his hands or clothes were visible. Assuming the murderer to be such a person as I have just described, he would be solitary and eccentric in his habits, also he is most likely to be a man without regular occupation, but with some small income or pension. He is possibly living among respectable persons who have some knowledge of his character and habits and who may have grounds for suspicion that he isn't quite right in his mind at times. Such persons would probably be unwilling to communicate suspicions to the Police for fear of trouble or notoriety, whereas if there were prospect of reward it might overcome their scruples.*" (Rumbelow, 1975, p. 138)

D. **1895: James McKeen Cattell** asked Columbia University students to answer questions, and indicate degree of confidence in their answer. Found much inaccuracy, setting stage for later research on eyewitness testimony.

E. **1896: Albert Von Schrenk-Notzing** testified at German murder trial re effects of pre-trial publicity. Based on research on memory and suggestibility, Schrenk-Notzing argued that witnesses would not distinguish what they saw and what was reported in the press.

F. **1901: William Stern** had students look at picture for 45 seconds and recall what happening in it after varying intervals. Recall generally inaccurate, more so with increasing delay. Recall most incorrect if asked leading question, e.g., "Did you see the man with the knife?". Recall accuracy decreased if strong emotions involved. Later founded first academic journal devoted to applied psychology.

G. 1908: **Hugo Munsterberg** (called first forensic psychologist) publishes "On the Witness Stand".
   1. Studied with Wundt at Leipzg. Moved to Harvard in 1892 and set up laboratory to study application of psychology to law.
   2. Quote: '*The lawyer and the judge and the juryman are sure that they do not need the experimental psychologist. They do not wish to see that in this field preëminently applied experimental psychology has made strong strides, led by Binet, Stern, Lipmann, Jung, Wertheimer, Gross, Sommer, Aschaffenburg, and other scholars. They go on thinking that their legal instinct and their common sense supplies them with all that is needed and somewhat more; and if the time is ever to come when even the jurist is to show some concession to the spirit of modern psychology, public opinion will have to exert some pressure. ... the following popular sketches ... deal essentially with the mind of the witness on the witness stand; only the last, on the prevention of crime, takes another direction. I have not touched so far the psychology of the attorney, of the judge, or of the jury problems which lend themselves to very interesting experimental treatment. ... my only purpose is to turn the attention of serious men to an absurdly neglected field which demands the full attention of the social community*' "
   3. Conducted research on witness memory, false confessions, the role of hypnosis in court. His research on discriminating sounds in quick succession played role in preparation for trial of Lee Harvey Oswald for assassination of President John F. Kennedy.

H. **1916: Lewis Terman** at Stanford, revised Alfred Binet's intelligence tests to create Stanford-Binet test. Used to assess intelligence of 30 police, firefighting applicants in San Jose, CA. Some years later, L. L. Thurstone used similar test in Detroit.

I. **1917: psychologist William Marston** (student of Munsterberg) found systolic blood pressure correlated with lying; discovery led to design of polygraph in early 1920s. Marston's testimony in 1923 (Frye vs. United States) established precedent for use of expert witnesses in courts. "The Federal Court of Appeals determined that a procedure, technique, or assessment must be generally accepted within its field in order to be used as evidence."

J. About 2,000 psychologists now belong to American Psychology-Law Society.

VII. Post-WWII Growth

A. Major growth in forensic psych in North America after Second World War. Before that, psychologists served as expert witnesses only in trials not seen as 'medical'. In 1940 case (People vs. Hawthorne) courts ruled that standard for expert witnesses was extent of knowledge of subject, not whether or not witness had medical degree.

B. In 1954 Brown vs. Board of Education case which overturned segregated schools, several psychologists testified for both plaintiffs and defendants. Later, courts supported psychologists serving as mental illness experts in case of Jenkins vs. United States (1962).

C. 2001, American Psychological Association officially recognized forensic psychology as specialization within psychology. Many forensic psychology journals available:
   1. American Journal of Forensic Psychology
   2. Behavioral Sciences & the Law
   3. Criminal Justice and Behavior
   4. Journal of Forensic Psychology Practice
   5. Law and Psychology Review
   6. Psychology, Crime, & Law
   7. Psychology, Public Policy, and Law

VIII. What do forensic psychologists do?

A. In civil cases:
   1. Child custody cases: Assess competence of parents, wishes of child, and best choice for the child's optimal development.
   2. Assess competence for civil purposes (e.g., changing will, living on one's own, etc.)
   3. Testify re psychological harm, or deception, in injury, workman's compensation cases

B. In criminal cases:
   1. For violent and especially serial crime (rape, murder), provide offender profile.
   2. When subject identified or arrested, assess for mental illness, and/or for fitness to stand trial.
   3. At trial:
      a. In U.S., but no so much in Canada, assist sides in selecting & evaluating the jury
      b. Assist defense in preparing defendant for best appearance before court

       c. Testimony on current mental state, or mental state at time of crime (usually forensic psychiatry)
       d. Expert testimony of the nature and validity of evidence:
          (1) Witness testimony
          (2) Effects of methods of interrogation
          (3) Validity of any methods of deception detection used (e.g., lie detection)
    4. At sentencing:
       a. Expert testimony about any mitigating psychological factors that might reduce (or enhance) the defendant's sentence or reflect on his or her chances for rehabilitation.
       b. Testimony about risks of violence or recidivism
    5. In prison:
       a. Interviewing for forensic information
       b. Involvement in treatment of offenders
    6. At parole hearings: Testimony about mental state and chances for re-offending or successful integration into society

C. In academia:
    1. Whether they appear in court or ever evaluate individuals psychologically, many experimental psychologists consider themselves 'forensic' because they conduct research in areas directly relevant to the law:
       a. Detecting truth and deception in individuals
       b. Memory and the factors that affect eyewitness testimony
       c. Jury deliberations and dynamics
       d. Offender characteristics
       e. Etc.

D. Where do forensic psychologists work?
    1. In private practice as a clinician
    2. As private or part-time consultant to attorneys for plaintiff/crown or defendant in criminal or civil action.
    3. In mental health unit, hospital in a clinical capacity
    4. In academia purely as a researcher:
    5. For a law enforcement agency:
       a. FBI in U.S.: Behavioral Analysis Unit (though few, and none have job as 'profiler')
       b. RCMP, CSIS

IX. Subfields of forensic psychology.

   A. **Clinical-Forensic Psychology**: Similar to clinical psychology, though not all clients suffering from a mental disorder. (See reference #5, at bottom)
      1. Assess and treat mental illness in those involved in criminal justice system (usually convicts). Assessment and treatment in prison, or in private office.
      2. Serve as expert witness in criminal or civil cases for either side
         a. Civil competence re decisions or actions (e.g., writing or altering a will).
         b. Psychological harm in workman's compensation cases
         c. Child custody cases (e.g., re which parent should be child's legal guardian.)
      3. May combine in-court activities with research on forensic issues; e.g., the relative effectiveness of different therapies for convicts.
      4. Career path typically requires a doctorate in psychology, and clinical internship in order to become a registered Psychologist in Ontario.

   B. **Developmental Psychology**: Deals with issues that relate law to person's age, e.g. juvenile offenders or children in custody cases. Focus on policy rather than treatment.
      1. Like clinical-forensic psychologists, may testify re child's ability to make life decisions (e.g., rejecting medical treatment, having abortion, which parent they want to live with.)
      2. May testify about juvenile's ability to understand legal proceedings, to give testimony, or how specific sentence options will affect the juvenile.
      3. In abuse cases, may testify re how abuse might affect individual in adulthood.
      4. May give evidence re ability of elderly person to manage their affairs, and need for guardianship in such cases.
      5. Psychologists in this area are typically academically employed, and may do research on all the areas in which they give testimony.

   C. **Social Psychology**: Primarily concerned with how jurors interact and arrive at decisions.
      1. Consult with attorneys, courts, and agencies re witness credibility, jury selection (though not in Canada), and juror decision making influences.
      2. Academic social psychologists show how social factors influence (e.g.) judge and jury decisions, credibility of witnesses, how memory affected by social factors, etc.

   D. **Cognitive Psychology**: Similar in relationship to forensic as social field, this area more interested in how people make decisions in legal cases. What are judges and juries thinking, and why? Special interest in memory issues, esp eyewitness identification and recovered memories..

   E. **Criminal Investigative Psychology**: These are the folks who are interested in police psychology, criminal profiling and psychological autopsies.
      1. Profiler creates psychological profile of suspect based on knowledge of motivation, mental illness, and behavior. Profilers examine actions at the crime scene, or behavior towards victim. Profiler can compare profile of UNSUB (Unknown SUBject) to the characteristics of other criminals or patients.
      2. Police psychologists work most closely with officers and victims as opposed to

criminals. May help evaluate police applicants, do counseling for both police and victims. Must have experience with psychological assessment and testing as well as counselling.

References

Bartol, C. R., & Bartol, A. M. (2005) History of Forensic Psychology. In I. B. Weiner & A. K. Hess (Ed.), *The Handbook of Forensic Psychology* (pp.1-27). Hoboken, NJ: Wiley.

Baskin, I. & Sommers, D. (2010). The influence of forensic evidence on the case outcomes of homicide incidents. *Journal of Criminal Justice*, 36(6), 1141-1149.

Cattell, J. M. (1895). Measurements of the accuracy of recollection. *Science*, 2, 761–766.

National Research Council (2011). *Reference Manual on Scientific Evidence, 3rd Edition.* National Academies Press, Washington, D.C.

Office of the Inspector General (2006). A Review of the FBI's Handling of the Brandon Mayfield Case. Washington, D.C.

Schwartz, A. (2005). A Systemic Challenge to the Reliability and Admissibility of Firearms and Toolmark Identification. *The Columbia Science and Technology Law Review.* Val. VI, pp. 1-42.

Stern, L.W. (1939). The psychology of testimony. *Journal of Abnormal and Social Psychology*, 40, 3–20.

Shelton, D.E., Kim, Y.S & Barak, G. (2007). A Study of Juror Expectations and Demands Concerning Scientific Evidence: Does the "CSI Effect" Exist? *Vanderbilt Journal of Entertainment and Technology Law*, 9(2), 331-368.

Suggested Links

http://www.wcupa.edu/_ACADEMICS/sch_cas.psy/Career_Paths/Forensic/Career08.htm
West Chester University's Department of Psychology. In Pennsylvania. Has excellent information about careers in forensic psychology.

http://www.crimezzz.net/forensic_history/index.htm . An interesting timeline of events in forensic science, and seemingly accurate, though not by someone active in the field professionally.

http://ww2.csfs.ca/. Canadian Society of Forensic Science

http://www.abfp.com/ American Board of Forensic Psychology. Provide certification in forensic psychology in the U.S.

http://www.aafp.ws/ American Academy of Forensic Psychology. The academic side of forensic psychology in the U.S. Provides workshops in forensic issues.

http://www.ap-ls.org/ American Psychology-Law Society. A division of the American Psychological Association. Has a small number of useful links: to information about careers in forensic psychology; to a list of references on eyewitness testimony, and the AP-LS News Bulletin.

http://www.crimelibrary.com/index.html. From CourtTV. Has some interesting links (under "The Criminal Mind") that are relevant to forensic psychology.

http://www.fbi.gov/hq/isd/cirg/ncavc.htm#bau. The FBI's Behavioral Analysis Unit, and several programs NCAVC and VICAP have links on this page. Also check the link at the left on "Reports and Publications"

Interview & Interrogation Procedures

I. Introduction

    A. We have all seen police interviews and interrogations on television and in the movies. Most are not that distant from the procedures that are actually used by police interviewers, though most contain a number of errors from the psychological - and occasionally from the legal - point of view.

    B. Will distinguish between:
        1. **Interviews** - with witnesses and other 'persons of interest' in a crime investigation designed to elicit information about the events and individuals involved in a crime, whether as perpetrators, witnesses, or victims. An interview is a relatively neutral term. Non-confrontational.
        2. **Interrogation** - with individuals thought likely to be perpetrators of a crime, or those who have assisted a perpetrator before and after a crime. Although designed to elicit information, may be more accusatory, and designed to elicit inculpatory information, including confessions.

II. Interview techniques:

    A. General agreement about procedures that make for good investigative interview. Seldom see one on TV or in movies:
        1. Most police officers not trained in appropriate interview techniques
        2. Appropriate interview techniques not often used in practice.
        3. A good interview makes terrible viewing

    B. Good investigative interview protocol has the following characteristics:
        1. **Establishment of rapport between the interviewer and interviewee**: Research shows that interviewee who is at ease in interview will provide more info - especially if the topic is sensitive or traumatic. Little research on how best to build rapport, but three aims or rapport-building:
            a. Interviewee does much more talking than the interviewer
            b. Interviewer conveys understanding and acceptance in a non-judgmental and non-coercive manner.
            c. Interviewer creates relaxed and informal context
        2. **Interviewee understand 'rules' of the interview:** In forensic interviews, interviewee is expert, and more detail is required from interviewee than usual. Interviewee must understand this, and mut be instructed:
            a. Hold no information back, even if it seems irrelevant, or if interviewee assumes interviewer already knows it (like stream-of-consciousness psychoanalysis)

b. Avoid guessing, assuming, or making up information
   c. Ask if a question is unclear
   d. Correct the interviewer if he or she misunderstands or misinterprets information
   e. Use any language that is comfortable, including profanity
   f. Repeated questions do not mean that the first answer was wrong or inappropriate
3. **Open-ended questioning style:** Whenever possible, request free-narrative response from interviewee.
   a. Open-ended questioning allows interviewee to work at his/her own pace, to collect thoughts before responding. It therefore promotes more elaborate memory retrieval.
   b. Answers to open-ended questions usually more accurate than responses to specific questions (**Lipton, 1977; Fisher et al, 2000; Philips et al, 1999**)
   c. Specific follow-up questions are more prone to elicit erroneous info than are the open-ended questions.
   d. Specific questions may lead interviewer to underestimate interviewee's language limitations, especially if he or she uses strategies to hide those limitations (e.g., by repeating words or phrases used by the interviewer, giving answers even when questions not understood, etc.).
4. **Willingness to explore alternative hypotheses:** Interviewer should not assume that preconceptions about case are correct. This could contaminate their interview, and lead to bias. Biased interviewers:
   a. Tend to report content of interviews such that it is more consistent with their bias.
   b. Tend to overlook important information that is inconsistent with their bias.
   c. Tend to distort witness account by asking misleading (closed) questions that reinforce the interviewer's views of event.
   d. Any misinformation provided by a biased interviewer distorts later testimony of the interviewee, even if the erroneous information was rejected by him or her (**Warren, Hulse-Trotter, & Tubbs, 1991**)

C. Where are we now?
1. Research shows most interviewers use short-answer questions with few pauses, many leading and closed-ended questions. Certainly see **THIS** on TV!
   a. Trend is international (reported in U.S., U.K, Australia, Israel, Sweden)
   b. Trend occurs whether interviewees are adults or children
   c. Trend occurs across different interviewer types (e.g., police, social workers)
2. Research finds **no relationship** between knowing best practices and actually using them in an interview.
3. Only positive findings concern higher proportion of open-ended questions by interviewers after training on the Cognitive Interview

D. **Compo, Gregory & Fisher (2012)**: Examined whether US police investigators adhere to nationally published guidelines when interviewing witnesses and victims of crime.
   1. Sample of audiotaped real-world witness interviews from 26 South Florida investigators analyzed, indicating that investigators rarely use recommended 'positive' interviewing techniques (e.g. rapport building or context reinstatement) while using many 'negative' techniques (e.g. interrupting the witness or using complex questions).
   2. Mean % and number of question types:
      a. Open-ended narrative : 6.53 /interview (10.8%)
      b. Closed-ended: 16.2 /interview (25.7%)
      c. Multiple-choice 2.7 /interview (4%)
      d. Yes/No 39 /interview (59%)
   3. Positive and negative techniques used (per interview):
      a. Positive:
         (1) Rapport-building (1.78)
         (2) Long pauses (5.04)
      b. Negative:
         (1) negative rapport-building (1.13)
         (2) interruptions (5.67)
         (3) suggesting leading questions (5.87)
         (4) multiple questions (3.87)
   4. Data suggest that national US recommendations on witness interviewing have not been translated into real world interviewing practice by the investigators surveyed.

E. **Wright & Alison (2004)**: Analyzed 19 Canadian police interviews with adult witnesses.
   1. Found that several interviewing strategies used were counter to recommendations in the literature; interviewers:
      a. Interrupted the witness more than necessary
      b. Seldom used 0cognitive techniques to enhance memory recall
      c. Asked many more closed-ended than open-ended questions.
   2. Also looked at sequence of questions and found that officers first "help" witness construct event, then asked sequence of "yes/no" questions, seek to confirm that account.
   3. Argued that officers begin with assumptions about the nature of the event and seek to confirm it.

III. The **cognitive interview**: Designed for adults who want to be helpful in recalling info, but need assistance.

A. **Geiselman (2012)**: Eight stages of the Cognitive Interview for suspects (CIS)
   1. **Rapport/Introduction**: Interviewer presents self as ordinary person, not an authority figure. Interviewer creates rapport with interviewee by showing interest in interviewee, and engaging on casual conversation. Interviewer observe interviewee's demeanor, the better to detect any later changes in demeanor that might signal deception.

2. **Narrative:** Interviewer instructs interviewee to be as detailed as possible, and to take as much time as needed to do that. Transfer of control to interviewee ("You were there, I was not, so I am relying on you for the information I need."); and reconstruction of the context of the event to be described. Interviewer uses prompts to keep narrative going as long as possible ("Really... tell me more about that."), since more detail allows more opportunities to detect inconsistencies in the story.
3. **Drawing/Sketch:** Interviewer asks for illustration of story from the narrative stage, ostensibly to clarify the narrative for greater understanding by the interviewer, and as a chance for interviewee to recall additional information.
4. **Follow-up, Open-Ended Questions:** The follow-up, open-ended questions should be presented as information-gathering rather than as confrontational to maintain the momentum toward generating more information from the suspect. Liars typically answer these questions with minimal elaboration without offering much that is new (19). Look for changes in body language from the rapport stage as well as for the more reliable indicators of deception (leakage)."
5. **Reverse-Order Technique**: When all scenes from narrative have been covered in follow-up questions, ask for recall of events in reverse order. Research shows that liars have great difficulty telling fabricated stories backward, since it increases cognitive load, and liar's cognitive resources already stretched thin in maintain consistency of false story.
6. **Challenge:** Confront interviewee re any inconsistencies, incriminating statements, and/or external incriminating evidence. Interviewer should remain soft-spoken, respectful. (Think of Columbo's technique in the old TV series.) Research suggests weakest evidence should be presented first. Research indicates that truthful interviewees tend to offer more information in response to challenge, while liars tend not to.
7. **Review**. Presented as a chance for interviewee to correct any inaccuracies and recall additional facts. With suspects, interviewer should change some non-incriminating element of the story to see if interviewee spontaneously corrects the inaccuracy. Liars tend not to, but quickly agree with the review, to end the interview as soon as possible.
8. **Closure**: Thank interviewee for cooperation if they seemed truthful. If not, interviewer indicates disappointed and disrespected.

B. **Reinstate Context**: Ask interviewee to think back to original event, recalling physical (time of day, workspace, etc.) as well as emotional context (rushed, bored, etc.)

C. **Report everything**: As in usual interview, interviewee encouraged not to hold back any information, even if it seems unimportant, or if subject thinks the interviewer knows already.

D. **Varied order of retrieval**: Ask for recall in reverse order, or starting from the middle and working to either end.

E. **Change perspectives**: Imagine scene from another perspective (e.g., security camera, another participant, etc.) This notion has been criticized since witness did not see event from other perspectives, and this instruction might induce false recollections.

F. Techniques:

1. Focused concentration: Interviewee may close eyes to minimize distractions; interviewer avoids interrupting or other intrusions to the sessions.
2. Extensive retrieval: Interviewer does not let interviewee stop after a cursory search of memory, but encourages multiple attempts.

G. Use of hypnosis to aid recall
   1. **Steblay & Bothwell (1994)** conducted meta-analytic review of 24 studies and concluded:
      a. Overall, recall accuracy less accurate for hypnotized subjects.
      b. Hypnotized subjects made more intrusion of uncued errors, and had higher levels of pseudomemory.
      c. Hypnotized subjects had higher confidence in recall accuracy compared to nonhypnotized subjects.
   2. Increases the quantity of recall, perhaps by lowering threshold for reporting, or by increasing confidence in one's memory: Number of correctly recalled details increases.
   3. Quality of recall is not improved: Number of incorrect answers also increases. **(Dywan & Bowers, 1983)**
   4. Subjects confidence in accuracy is not different for correct and incorrect recall
   5. Typically not allowed in U.S. or Canadian courts because of known problems

IV. Interrogation:

A. Typically take place in coercive environment. U.S. movies from 1930s, 1940s often show suspects being coerced or beaten into confessing. Rough tactics still occur today, though less common in North America.
   1. Traditionally, and occasionally today, info including confessions extracted using physical violence, or the threat of same
   2. Police my lie, make promises of leniency, or make veiled threats against suspect or his family, etc.
   3. **Inbau et al** (2001) states that suspects **must** be tricked into confessing.

B. The **Reid Model** of interrogation (based on description of the late John E. Reid's model as described in **Inbau et al, 2001**).
   1. Based on three fundamental assumptions: **(Inbau et al, 2004)**
      a. *"Many criminal cases, even when investigated by the best qualified police departments, are capable of solution only by means of an admission or confession from the guilty individual or upon the basis of information obtained from the questioning of other criminal suspects."* (Forensic evidence gathered in only about 10% of criminal cases, and only about half of that evidence is analyzed. (Inbau et al, 2004, p, xii)
      b. *"Criminal offenders, except those caught in the commission of their crimes, ordinarily will not admit their guilt unless questioned under conditions of privacy and for a period of perhaps several hours."*

    c. *"In dealing with criminal offenders, and consequently also with criminal suspects who may actually be innocent, the investigator must of necessity employ less refined methods than are considered appropriate for the transaction of ordinary, everyday affairs by and between law-abiding citizens."*

2. A three-part process:
   a. Part I: Gather evidence, interview victim and witnesses
   b. Part II: Non-accusatory interview of suspect to assess possible guilt
   c. Part III: Accusatory interview of suspect if he/she is judged to be guilty
3. Part III: Nine-step process
   a. **Direct Positive Confrontation**: Indicate strong confidence in the suspect's guilt by direct statement, e.g.: "Our investigation clearly indicates that you did this crime." Use evidence file or other props (such as videotape). Establish that the purpose of the interrogation is to establish the circumstances behind what happened; trying to find out what kind of person the suspect is; trying to establish the extent and frequency of his involvement; and/or, to find something about the suspect that can be used as the basis for a compliment.
   b. **Theme Development**: Present the suspect with reasons and excuses that morally (but not legally) excuse suspect's behavior. This allows suspect to reduce the perceived consequences associated with crime, both legal and personal (e.g., self-esteem). Present all of this in compassionate and understanding manner.
   c. **Handling Denials**, Prevent suspect from engaging in denials though (e.g.) verbal phrases ("Jim, wait just a minute") and nonverbal gestures (e.g., turning head away; using hand in a stop gesture).
   (http://www.reid.com/educational_info/critictechniquedefend.html)
   d. **Overcome suspect's objections to charges** until he becomes quiet and withdrawn. Often done by agreeing with suspect, and then mentioning other possible reasons for his behavior.
   e. **Reduce psychological distance between suspect and interrogator**, e.g., by moving closer. "Another tactic that may be used to illustrate this emotional closeness would be to touch the suspect with a hand on their arm, for example, to extend sympathy and understanding, not to exert authority over the suspect."
   f. **Interrogator shows sympathy and understanding**, urging suspect to confess.
   g. **Alternative question: Interrogator offers suspect face-saving explanations for crime**, easing self-incrimination. Offer the suspect two incriminating choices; one is face-saving:
      (1) "If this was planned out months in advance that tells me you have a criminal mind and are capable of committing much more serious crimes than this and I think that's despicable. But if this happened on the spur of the moment I can understand that. Did you plan this out for months in advance, or did it pretty much just happen on the spur of the moment?" "
      (2) "If you want people to believe that you are greedy and spent this money on fancy things for yourself then my advise to you is to say nothing. But if you really needed this money for your family that would be important to know. Did you spend the money on yourself, or did you use it to help out your family?"
      (3) "If this was not your idea and somebody talked you into it that would be very important for me to include in my report. On the other hand, if this whole thing

was your idea and you were the mastermind behind it, you're doing the right thing by saying nothing. Was this whole thing your idea, or did you get talked into it?"
   h. **Once suspect admits responsibility, interrogator develops this into full confession**: interrogator verbally develop details of the crime, and converts verbal confession into a recorded or written document. Develop the admission into a confession by asking open-ended questions that allow the suspect to provide details of his criminal behavior, e.g., "Then what happened?"; "Where did you go next?"; "What did you do after that?" Ideally, want to get corroboration of details of the crime known to investigators, but not released publicly, or information about the crime that was unknown until the confession, but can be independently verified.
   i. **Interrogator convinces suspect to write and sign full confession**.
4. Model based on the idea that suspects don't confess because they fear consequence of confessing. Goal is to make consequences of confession seem less than continuing guilt and anxiety.
5. Used 'good cop/bad cop' techniques:
   a. Good cop (**minimization**): designed to "lull the suspect into a false sense of security" (Kassin, 1997, p. 223). Include the use of sympathy, excuses, and justifications.
   b. Bad cop (**maximization**): designed to "intimidate a suspect believed to be guilty" (Kassin, 1997, p. 223). Can be done by exaggerating seriousness of offense, and deceiving suspect about evidence available to police.
   c. Courts often discount or disallow evidence obtained through direct of implied threat (maximization), but are more open to confessions based on minimization.
      (1) See also Kassin & McNall !991);
6. Problems with Reid model of interrogation: Lots of research on the Reid model. (See Ekman & O'Sullivan, 1991; Kassin, Goldstein & Savitsky, 2003; Ofshe & Leo, 1997). Three main problems:
   a. **Difficulty in detecting deception during initial interrogation** (Ekman & O'Sullivan, 1991). General tenor of research findings is that even after special training (Köhnken, 1987) investigators are not very good at detecting deception in suspects. (But see Porter, Woodworth & Birt, 2000). See chapter 5 in text.
   b. **Biases in interrogation and treatment when subject believed to be guilty (Kassin, Goldstein & Savitsky, 2003)**. Authors compared interrogation techniques in mock situation where some interrogators believed Ss guilty, others believed them innocent. Found that interrogators with guilty expectations:
      (1) Asked more questions indicating belief in Ss guilt.
      (2) Used a higher frequency of interrogation techniques, especially early in the interrogation.
      (3) Judged more Ss to be guilty, whether they were or not.
      (4) Interrogators indicated that they exerted more pressure for confession when they believed S to be guilty than when believed to be innocent.
      (5) Ss had fairly accurate perceptions of the interrogators's behavior
      (6) Neutral observers saw interrogators with guilty expectations as more coercive, especially against innocent subjects, and viewed Ss in guilty condition as more defensive.

c. **Possibility that coercive interrogation techniques will lead to false confessions** (Ofshe & Leo, 1997). Reid responds to these criticisms on his web page.
7. **Horgan et al (2012):** Examined whether one can distinguish between minimization and maximization techniques that do or do not influence a suspect's perceptions of the consequences of confessing using mock suspects rating statements that might be made by interrogator if they were being interrogated
   a. Techniques that vary perceived consequences of confessing:
      (1) Maximization:
         (a) exaggerating consequences of actions
         (b) pitting con-conspirators against each other
      (2) Minimization:
         (a) stressing the benefits of cooperation
         (b) downplaying the consequences of actions
         (c) providing face-saving excuses for actions
         (d) minimizing the seriousness of the offense
   b. Techniques that do **NOT** vary the perceived consequences of confessing:
      (1) Maximization:
         (a) assuming an unfriendly demeanor
         (b) expressing a firm belief in the suspect's guilt
      (2) Minimization:
         (a) expressing sympathy for the suspect
         (b) assuming a friendly demeanor
         (c) flattering, boosting suspect's ego
         (d) appealing to suspect's conscience
   c. Results indicate that techniques that manipulate the perceived consequences of confessing influence both the decision to confess and the diagnostic value of confession evidence.

V. Interrogation Practices and the Courts

   A. Acceptance of confessions in court depends on judge's decision that confession was voluntary, and that the suspect was competent to make the statement at the time.
      1. Kassin, 1997, p. 221: "A confession is typically excluded if it was elicited by brute force; prolonged isolation; deprivation of food or sleep; threats of harm or punishment; promises of immunity or leniency; or, barring exceptional circumstances, without notifying the suspect of his or her constitutional rights"
      2. But in U.S. and Canada, confessions elicited under more subtle forms of coercion are often accepted.

VI. Recent changes to interrogation procedures:

A. In England (see Gudjonsson, 1992a), the courts have restricted many aspects of the Reid interrogation model. Interrogations now guided by PEACE model: **Planning & Preparation; Engage & Explain; Account; Closure; Evaluation**

B. **Planning and preparation**:
   1. Crucial elements of good planning:
      a. Define the aims and objectives of the interview
      b. Obtain background information on the incident, and on interviewee
      c. Recognize and understand the points to prove
      d. Assess available evidence and where it was obtained
      e. Assess the evidence needed, and how it can be obtained
      f. Set the stage for the interview (exhibits, location, equipment, seating, etc.)
   2. Disadvantages of poor preparation and planning include:
      a. Overlooking important evidence
      b. Not identifying inconsistencies and lies
      c. Needing unnecessary breaks to get further information
      d. Needing additional interviews with the same person
      e. Losing control of the interview.

C. **Engage and Explain (establishing rapport)**: Police interview is nerve-wracking for most people; 'warm-up' period needed to settle interviewee down. Interviewers need to engage interviewee so relaxed, cooperative relationship exists throughout the interview.
   1. Crucial steps in this process:
      a. Create a good impression from the start by being courteous, polite, and understanding.
      b. Treat the interviewee as an individual by determining and addressing the interviewee's needs and concerns.
      c. Understand the feelings of the person being interviewed. Empathize while remaining objective.
      d. Explain the reason for the interview: Emphasize the importance of the interviewee's knowledge in assisting the investigation.
      e. Outline and explain the interview procedures.
         (1) Describe format of interview: Interviewee will be invited to give an account of the event; to clarify it through supplementary questions; to comment on matters not been covered or not fully explained; the interviewer will provide a verbal summary from time to time to verify his interpretation.
         (2) During this stage, the interviewer should take their time and use plain English. They should show consideration by:
            (a) being concerned for interviewees' welfare (do they want water, or to know where the toilet is)
            (b) asking how they want to be addressed (e.g., Christian name, or by title and surname)
            (c) checking how long they've got (e.g., do witnesses have to be somewhere by a certain time, or are they on a parking meter)

             (d) giving reassurance if the person seems nervous of the process.
        f.   Interviewers should let interviewee know:
             (1) that their statement is important so they should report completely and without omissions
             (2) not to omit info even if they believe it is irrelevant or known to the interviewer
             (3) that they are the experts, and so will do most of the work in the interview
             (4) that they should feel free to alert the interviewer if:
                 (a) they do not understand a question
                 (b) they do not know the answer to a question
                 (c) the interviewer misunderstands what the interviewee has said
                 (d) the interviewer asks a leading or inappropriate question
             (5) that they may be asked additional questions to clarify their account
        g.   The interviewer can use this time to assess the interviewee's communication abilities and modify his or her language to match that of the interviewee.
        h.   Courtesy, respect and professionalism will help even when interviewees are evasive, deceptive, unwilling to give information, or openly hostile.

   D. **Account**: Interviewer obtains interviewee's account of events. The three main steps are:
      1. Obtain interviewee's uninterrupted account
      2. Expand and clarify that account
      3. If necessary (e.g., with suspects) challenge interviewee's account

   E. **Closure**: Purpose of closure is to:
      1. Ensure mutual understanding of interview by reviewing and summarizing interviewee's account
      2. Verify that everything has been covered by checking that interviewee has given all information he or she is able and willing to provide
      3. Explain what will happen next, by giving the interviewee appropriate information on the next stages of the process e.g., telling witnesses whether or not they will have to attend court.

   F. **Evaluation**: A wrap-up stage in which the interviewer:
      1. Determines whether the aims and objectives of the interview have been met
      2. Reviews the investigation in the light of information obtained during the interview
      3. Reflects upon how well he or she conducted the interview and considers what improvements could be made in future

   G. Little research to evaluate the PEACE model, but some indications are that reduced coercion does not necessarily lead to fewer confessions (Meissner & Russano, 2003)
      1. Both in U.S. and Canada, use of video to record confessions becoming more common.

VII. False Confessions

   A. The big concern in interviews and interrogation is getting incorrect information, with a focus on detecting deception (our next topic). We place most emphasis on subjects who deceive by eliminating self-incriminating information, but another concern is individuals who falsely incriminate themselves.

   B. In his 1908 book 'On The Witness Stand: Essays on Psychology And Crime', Hugo Munsterberg included an essay on 'Untrue Confessions'. Munsterberg said:

   C. Useful to distinguish between:
      1. **False confession**: The contents of the statement of guilt are false
      2. **Retracted confession**: Contents may be false or true, confessor now says they are false.
      3. **Disputed confession**: Legal issues about whether they can be entered as evidence.

   D. No indication of how frequent false confessions are, but they have at least two undesirable consequences, so it is important to understand why they happen, and what can be done to avoid and detect them.
      1. Jurors give high weight to confessions (**Kassin & Neumann, 1997**)
      2. Possible false conviction: In mock jury situations, even an obviously false and coerced confession increases the probability of conviction. (**Kassin & Sukel, 1997**)
      3. Diversion of police efforts, and possible escape of true offender

   E. Types of false confessions:
      1. **Voluntary, false confessions**: Without coercion, suspect confesses to crime he or she did not commit. There may be many reasons for false confessions:
         a. Desire for notoriety
         b. Mental illness and deficient contact with reality
         c. Psychological need to atone for other real or imagined sins or offenses
         d. Protecting someone else from punishment
      2. **Coerced-compliant false confessions**: Confession made under duress from real or perceived threats from interrogator. Confessions designed to:
         a. End the interrogation process
         b. Earn rewards promised by interrogator
         c. Avoid threatened punishments
         d. Critics argue that the Reid model of interrogation is especially likely to lead to coerced-compliant confessions
      3. **Coerced-internalized false confessions**: Confession under pressure, but individual comes to believe its truth.
         a. **Gudjonsson (1992b)** argues for several vulnerability factors:
            (1) History of substance abuse "of some other interference with brain function"
            (2) Inability to detect differences between their own experiences and what has been suggested to them
            (3) Anxiety, confusion, feelings of guilt
         b. The Paul Ingram case (Ofshe & Watters, 1994)

F. **Russano et al (2005)**: Participants (and a confederate) work on problems, some collaboratively, some independently.
   1. In one condition, confederate convinces P to collaborate, against E's instructions.
   2. E accuses Ps of collaborating, based on supposed similarity of answers to problems.
      a. Minimization condition: "Interrogator instructed to lessen seriousness of offense by … express[ing] sympathy and concern, offer[ing] face-saving excuses (e.g., ``I'm sure you didn't realize what a big deal it was"), and suggest[ing] to participants that it was in their interest to cooperate by signing [confession].
      b. No-minimization condition: No statements made.
      c. Deal condition: "Ps were told that if they signed confession, then ``things could probably be settled pretty quickly." Ps assured they would receive credit for the day, but would have to return for another session without credit. … also told that if they did not sign statement, E would call professor into lab, and prof would handle things as he saw fit, with implication that consequences would be worse if prof involved.
      d. In no-deal condition, Ps told that whether they signed statement or not, E would call professor back and find out what to do next.
   3. If P denied allegation or hesitated to sign, E repeated interrogation script up to three more times. If participant still refused, interrogation terminated.
   4. Guilty persons more likely to confess than the innocent; Minimization and Deal increased both true and false confessions.

G. **Kassin et al (2005)**:
   1. Worked with 10 prison inmates who truthfully confessed to crimes, or made false confession concocted for the study.
   2. University students and police investigators watched or listened to these confessions, and had to judge which were true, and which were false, and how confident they were in their judgments.
   3. Students more accurate than police, and accuracy rates for audio only than with videotaped confessions.
   4. Police significantly more confident in their judgments and more likely to judge confessors guilty.
   5. In second study, participants told that half confessions were false. This reduced proportion of confessions judged as true, but did not increase accuracy or reduce confidence.

H. **Shaw & Porter (2015)** using suggestive memory-retrieval techniques, participants induced to generate criminal and noncriminal emotional false memories, which were compared to true memories of emotional events.
   1. After 3 interviews, 70% of participants had created false memories of committing a crime (theft, assault, or assault with a weapon) that led to police contact in early adolescence and had volunteered a detailed false account.
   2. Reported false memories of crime similar to false memories of noncriminal events and similar to true memory accounts in that all had the same complex descriptive and multisensory components.

VIII. Interrogative Suggestibility and Compliance

   A. Distinction between compliance and suggestibility:
      1. **Compliance** = "tendency to go along with demands made by people in authority, even though the person may not agree with them" (Gudjonsson, 1989)
      2. **Suggestibility** = "tendency to accept (i.e., internalize) information communicated during questioning" (Gudjonsson, 1984)

   B. **Gudjonsson Compliance Scale** (GCS) developed by Dr. Gisli Gudjonsson at the Institute of Psychiatry, London UK.
      1. 20 T/F questions assessing person's eagerness to please others, and desires to avoid conflict and confrontation
      2. Gudjonsson (1989) compared GSC scores for 'resisters' (non-confessors despite evidence of guilt) and 'false confessors' (those who retracted previous confessions. Found resisters had lower GSC scores.

   C. **Gudjonsson Suggestibility Scale** (GSS) developed by Dr. Gisli Gudjonsson at the Institute of Psychiatry, London UK. Assessed person's eagerness to please others, and desire to avoid conflict and confrontation
      1. 20 T/F questions based on story told to individual. 15 of the 20 questions are Suggestive in that they imply information about the story that is not true. The remaining 5 questions are non-suggestive.
      2. Administration involves:
         a. Story is read to the Subject
         b. Subject gives free recall of the contents of the story
         c. Subjects is asked the 20 GSS questions (gives Yield 1).
         d. Subject is given critical feedback on his responses.
         e. Subject is given the 20 questions again (giving Shift and Yield 2).

References

Dywan, J. & Bowers, K. The use of hypnosis to enhance recall. Science, 1983, 222, 184-185.

Ekman, P. & O'Sullivan, M. Who can catch a liar? *American Psychologist*, 1991, 46, 913-920.

Fisher, R.P. (1995). Interviewing Victims and Witnesses of Crime. Psychology, Public Policy & Law, 1995, 1(4), 732-764.

Fisher, R.P. & Geiselman, R.E. 1992: *Memory-enhancing techniques in Investigative Interviewing: The Cognitive Interview*. Springfield, IL, Thomas. [$219!!]

Fisher, R.P., Falkner, K.L., Trevisan, M., & McCauley, M.R. Adapting the cognitive interview to enhance long-term (35 years) recall of physical activities. Journal of Applied Psychology, 2000, 85, 180-189.

Geiselman, R.E. (2012). The Cognitive Interview for Suspects (CIS). *American Journal of Forensic Psychology*, 30(3).

Geiselman, R.E., Fisher, R.P. MacKinnon, D.P. & Holland, H.L. Eyewitness Memory Enhancement in the Police Interview: Cognitive Retrieval Mnemonics Versus Hypnosis. *Journal of Applied Psychology*, 1985, vol. 70, No. 2, p. 403

Gudjonsson, G.H. A new scale of interrogative suggestibility. *Personality and Individual Differences*, 1984, 5, 303-314.1984.

Gudjonsson, G.H. Compliance in an interrogation context: A new scale. *Personality and Individual Differences*, 1989, 10, 535-540.

Horgan, A. J., Russano, M. B., Meissner, C.A. & Evans, J.R. (2012): Minimization and maximization techniques: assessing the perceived consequences of confessing and confession diagnosticity, *Psychology, Crime & Law*, 18:1, 65-78.

Inbau, F.E., Reid, J.E., Buckley, J.P., & Jayne, B.C. Criminal Interrogation and Confessions (4th Ed.). Gaithersberg, MD. Aspen, 2001.

Kassin, S.M. The psychology of confession evidence. *American Psychologist*, 1997, 52, 221-233.

Kassin, S.M., Goldstein, C.C. & Savitsky, K. Behavioral confirmation in the interrogation room: On the dangers of presuming guilt. *Law and Human Behavior*, 2003, 27, 187-203.

Kassin & Kiechel (1996): The social psychology of false confessions: Compliance, internalization, and confabulation. *Psychological Science*, 1996, 7, 125-128.

Kassin, S.M. & McNall, K. Police interrogations and confessions: Communicating promises and threats by pragmatic implication. *Law and Human Behavior*, 1991, 15, 233-151.

Kassin, S.M. & Neumann, K. On the power of confession evidence: An experimental test of the fundamental difference hypothesis. *Law and Human Behavior*, 1997, 21, 469-484.

Kassin & Sukel (1997). Coerced confessions and the jury: An experimental test of the 'harmless error' rile. *Law and Human Behavior*, 1997, 21, 27-46.

Köhnken, G. Training police officers to detect deception eyewitness statements: Does it work? *Social Behavior*, 1987, 2, 1-17.

Koriat, A., Goldsmith, M., & Pansky, A. Toward a psychology of memory accuracy. *Annual Review of Psychology*, 2000, 51, 481-537.

Lipton, J.P. On the psychology of eyewitness testimony. *Journal of applied Psychology*, 1977, 62, pp. 90-93.

Meissner, C.A. & Russano, M.B. The psychology of interrogations and false confessions: Research and Recommendations. *The Canadian Journal of Police and Security Services: Practice, Policy, and Management*. 2001, 1, 53-64. [this issue not available online]

Memon, A. & Bull, R. The cognitive interview: Its origins, empirical support, evaluation and practical implications. *Journal of Community and Applied Social Psychology*, 1991, 1, 291-307. [this volume not available online]

Ofshe, R.J. & Leo, R.A. The social psychology of police interrogation: The theory and classification of true and false confessions. *Studies in Law, Politics, and Society*, 1997, 16, 189-251. [journal not available online; Elsevier only goes back to Vol. 21]

Ofshe, R. & Leo, R. The Consequences of False Confessions: Deprivation of Liberty and Miscarriages of Justice in the Age of Psychological Interrogation. *Journal of Criminal Law and Criminology*, 1998.

Ofshe, R.J. & Watters, E. (1994). *Making Monsters: False Memories, Psychotherapy, and Sexual Hysteria*. New York: Charles Scribner's.

Philips, M.R., Fisher, R.P., & Schwartz, B.L. (July, 1999). *Metacognitive control in eyewitness testimony*. Paper presented at the meeting of the Society for Applied Research in Memory and Cognition. Boulder, CO.

Porter, S., Woodworth, M. & Birt, A. Truth, lies, and videotape: An investigation of the ability of federal parole officers to detect deception. *Law and Human Behavior*, 2000, 27, 216-233.

John Reid website: http://www.reid.com/educational_info/critictechniquedefend.html

Schacter, D.L. The seven sins of memory. *American Psychologist*, 1999, 54, 182-203.

Schollum, Mary. Review of investigative interviewing. Wellington, NZ. Police National Headquarters, 2005. ISBN 0-477-10011-2

Schreiber Compo, N., Hyman Gregory, A. & Fisher, F. (2012): Interviewing behaviors in police investigators: a field study of a current US sample, *Psychology, Crime & Law*, 18:4, 359-375.

Shaw, J. & Porter, S. (2015) Constructing Rich False Memories of Committing Crime. *Psychological Science* OnlineFirst, published on January 14, 2015.

Steblay, N.M. & Bothwell, R.B. Evidence for hypnotically refreshed testimony: The view from the laboratory. *Law and Human Behavior*, 1994, 18, 635-651.

Wakefield, H. & Underwager, R. Coerced or nonvoluntary confessions. *Behavioral Sciences and the Law*, 1998, 16, 423-440.

Walsh, D. & Bull, R. (2015). Interviewing suspects: examining the association between skills, questioning, evidence disclosure, and interview outcomes. *Psychology, Crime & Law*, Vol. 21 (7), 661-680.

Walsh, D., & Bull, R. (2012). Examining rapport in investigative interviews with suspects: Does its building and maintenance work? *Journal of Police and Criminal Psychology*, 27(1), 73–84.

Warren, A.R., Hulse-Trotter, K. & Tubbs, E.C. Inducing resistance to suggestibility in children. *Law & Human Behavior*, 1991, 15, 273-285.

Wilson, C.J. & Powell, M.B. *A Guide to Interviewing Children: Essential Skills for Counsellors, Police, Lawyers, and Social Workers*. Crows Nest, New South Walles, Allen & Unwin, 2001.

Wright, A.M. & Alison, L. (2004) Questioning Sequences in Canadian Police Interviews: Constructing and Confirming the Course of Events? *Psychology, Crime & Law*, Vol. 10(2), pp. 137 - 154.

Deception Detection

I. Introduction:

　A. Everybody deceives and tell lies - and is told lies. usually of the 'white lie' variety: "No, those jeans don't make you look fat."; "It's not you, it's me."

　B. The issue in forensic psychology is how to tell when someone involved in a legal case is telling the truth. Usually this is in a criminal case, but the issue also arises in civil cases when injury, or mental and emotional states are involved. Issues of malingering (pretending to have an injury or mental disorder) come under this heading, but we will discuss them later, under assessment. In this section, we will be concerned with deception by witnesses and those directly involved in crime.

II. How good are we at detecting deception on our own?

　A. Most people believe that they can detect deception. Certainly it is a big part of the responsibility of law officers, lawyers, and jury members to assess the credibility of those they deal with. What cues would we use to detect deception?

　B. **Facial and body movement and language**:
　　1. Many different behavioral characteristics thought to be (and reported to be) associated with deception (see **Vrij, 1998**). Both lay people (e.g., college students) and 'experts' (e.g., police officers, prison guards, prisoners) agree on these :
　　　a. Avoiding or failing to make eye contact
　　　b. Higher levels of smiling and laughter
　　　c. higher rate of eye blinking
　　　d. Nervous fidgeting
　　　e. Higher level of gestures to illustrate story
　　　f. Increased movement of extremities (hands, fingers, legs, feet)
　　　g. Increased body and/or head movements, shifts in posture or position
　　　h. More shrugs
　　2. But meta-reviews of relationship between cues and lying (see **DePaulo et al, 2003**) show that none is reliable. DePaulo et al looked at more than 150 cues to deception, and found that:
　　　a. Liars less forthcoming than truth tellers
　　　b. Liars tell less compelling tales
　　　c. Liars make a more negative impression and are more tense.
　　　d. Deceptive stories include fewer ordinary imperfections and unusual contents.
　　　e. "Cues to deception were **more pronounced** when people were motivated to succeed, especially when the motivations were **identity relevant** rather than monetary or material. Cues to deception were also stronger when lies were about transgressions." (**DePaulo et al**, abstract):

3. **NRC (2003)**: National Research Council in U.S. did large review of all studies of deception (we'll refer to it repeatedly), and found that:
    a. *"The meta-analytic literature fails to identify any pattern of facial or body movement that generally signals deception. However, some studies designed to develop indicators based on these movements show some ability to discriminate lying from truth-telling. For example, Ekman and his colleagues studied lying or truth-telling under fairly strong motivational conditions about three different matters: emotions felt at the moment (Ekman et al., 1991), a strongly held belief, and whether money was taken (Frank and Ekman, 1997). These studies suggest that the right measures of facial and motion features can offer accuracy better than chance for the detection of deception from demeanor in somewhat realistic situations."* (Pp. 164)

C. **Linguistic analysis**: What about detecting deception from verbal behavior?
   1. Text lists some of the many verbal characteristics that have been reported (by one study or another) to be associated with deception:
      a. False accounts less detailed (some support for this)
      b. Frequent pauses, speech fillers ('ah', 'um')
      c. Speech errors (repeating words or sentences, failing to complete sentences, slips of the tongue)
      d. Changes (especially increases) in voice pitch
      e. Faster speaking rate
   2. **DePaulo et al (2003)** looked at more than 150 cues to deception, and found that:
      a. Liars less forthcoming than truth tellers [less detailed, as above]
      b. Liars tell less compelling tales
      c. Liars make a more negative impression and are more tense.
      d. Deceptive stories include fewer ordinary imperfections and unusual contents.
      e. "The verbal and vocal immediacy measure is based on raters' overall impressions of the degree to which the social actors seemed direct, relevant, clear, and personal. The nonverbal immediacy measure includes the set of nonverbal cues described by Mehrabian (1972) as indices of immediacy (e.g., interpersonal proximity, leaning and facing toward the other person). The verbal composite and the verbal and nonverbal composite both indicated that liars were less immediate than truth tellers (d −0.31 and −0.55, respectively). Liars used more linguistic constructions that seemed to distance themselves from their listeners or from the contents of their presentations, and they sounded more evasive, unclear, and impersonal. The nonverbal composite was only weakly (nonsignificantly) suggestive of the same conclusion (d −0.07)." (DePaolo et al, 2003, p. 93)
      f. "Cues to deception were **more pronounced** when people were motivated to succeed, especially when the motivations were **identity relevant** rather than monetary or material. Cues to deception were also stronger when lies were about transgressions." (**DePaulo et al**, abstract):
   3. **NRC (2003)**: "Several different aspects of language use associated with deception. Strongest associations = immediacy of expression (e.g., using active or passive voice, affirmations or negations), observers' subjective impressions more strongly correlated with deception than objective measures tested (**DePaulo et al.**, 2003).

a. **Smith (2001)** evaluated scientific content analysis, (Sapir, 1987), using statements by criminal suspects later determined to be false or true.
    (1) Trained policemen correctly detected 80 percent of truthful statements and 75 percent of deceptive statements,
    (2) Untrained experienced policemen just as accurate.
    (3) Study cannot determine whether examiners making judgments based on their own experience rather than by using technique." (Pp. 165)
4. **NRC (2003) argued that** the analysis of language and facial and body movement might be useful in deception detection, especially when:
    a. Lies have high personal relevance
    b. Stakes are high
    c. Liar knows he or she is telling a lie when it is being told
    d. Before liar has chance to rehearse the lie

D. Remember **Kassin et al (2005)**: "College students and police investigators watched or listened to 10 prison inmates confessing to crimes. Half the confessions were true accounts; half were false—concocted for the study. Consistent with much recent research, students were generally more accurate than police, and accuracy rates were higher among those presented with audiotaped than videotaped confessions. In addition, investigators were significantly more confident in their judgments and also prone to judge confessors guilty. To determine if police accuracy would increase if this guilty response bias were neutralized, participants in a second experiment were specifically informed that half the confessions were true and half were false. This manipulation eliminated the investigator response bias, but it did not increase accuracy or lower confidence. These findings are discussed for what they imply about the post-interrogation risks to innocent suspects who confess."

E. **Matsumoto & Hwang (2014)**: "examined differences in word usage between truth tellers and liars in a moderately high stakes, real-life scenario (mock crime) involving participants from four cultural/ethnic groups—European-Americans, Chinese, Hispanics and Middle Easterners.
    1. Used LIWC (Linguistic Inquiry & Word Count) to assess word usage in written answers and transcripts of interviews.
    2. Each participant produced a written statement and participated in an investigative interview; word usage in both was analyzed.
    3. Word usage differed between truth and lies in both written statements and investigative interviews; effect sizes were substantial.
        a. For written statement, word usage predicted truths from lies at 68.90% classification accuracy;
        b. For investigative interview, word usage predicted truths from lies at 71.10% accuracy.
        c. Ethnicity did not moderate these effects.
        d. In written statements: (69% correct classification; 15% FP; 16% FN)
            (1) liars used:
                (a) fewer words overall
                (b) fewer words related to money
                (c) fewer words related to sensory processes, negations, motion and time

(d) used more tentative words (contrary to previous findings), and more equivocation words
(e) used more words related to positive emotions (previous meta-analysis indicated that liars tend to use more words expressing both positive and negative emotion, Hauch et al., 2012.
(f) used fewer articles, and fewer words related to motion
e. In interviews, liars used: (71% correct classification; 19% FP; 10% FN)
(1) more words related to first (I) and third person (Shehe) references
(2) fewer words related to motion

F. Findings not always consistent; two studies reported negative findings on words related to complexity and pronoun use:
1. **Hancock et al., 2008**
2. **Toma & Hancock, 2012).**

III. Empirical tests of deception detecting ability.

A. **Ekman and O'Sullivan, 1991:** U.S. Secret Service agents averaged 64 percent correct judgments (chance = 50 percent.) Half achieved accuracy of 70% or more, and 30% scored above 80%. (**Ekman and O'Sullivan, 1991**).

B. **Ekman, O'Sullivan & Frank (1999)**: Showed videotapes of speakers making true or false statements to federal law enforcement officials, clinical psychologists, deception-interested clinical psychologists, academic psychologists, and judges.

Subjects (n)	Overall Accuracy%	Lie Accuracy	Truth Accuracy
Federal officers (23)	73	80	66.1
Sheriffs (43)	66.7	77.7	55.8
Federal judges (84)	62	60.9	63.1
Mixed Law Enforcement Officers (46)	50.8	47.8	53.9
Deception-interested clinical psychologists (107)	67.5	71	63.9
Regular clinical psychologists (209)	62.1	64.3	59.8
Academic Psychologists (125)	57.7	57	58.4
	Overall Mean		

C. **Porter, Woodworth, & Birt (2000)**: Canadian study showing that parole officers at 40% in detecting truth/lies from videotaped statement. Accuracy improved to 77% after detection deception workshop.

D. **Bond & DePaulo 2006)**. Condicted meta-analysis of 206 studies involving more than 24,000 judges. Concluded that, when people try to distinguish truth from lies in real time and with no special aids or training, they average 54% correct judgments.
   1. They are correct 47% of the time in detecting lies
   2. They are correct 61% of the time in identifying true statements.
   3. People more accurate judging audible rather than than visible lies.
   4. Found that liars motivated not evade detection are actually more likely to be detected.

E. Recent study (**Carter & Weber, 2010**) indicates that people higher in trust of others are better lie detectors than people who are less trusting.

F. **ten Brinke, Stimson & Carney (2014)** found evidence from short video clips that conscious assessments of deception are at or below chance, while unconscious, implicit measures of deception detection were significant.
   1. But in a rebuttal to this study, **Levine & Bond (2014)** argues that their 2003 meta-analysis of deception detection showed that direct measures of deception detection produce better accuracy than indirect measures, and that ten Brinke et al's direct accuracy results are lower than those typically obtained in the literature.

G. **Klein & Epley (2015)** found that group discussion improves lie detection. Had individuals or groups of 3 individuals judge from video clips whether the speaker was lying or telling the truth. They also worked with 'nominal' groups, combining the judgments of people who had made their judgments of truth or deception separately.
   1. Speakers were giving responses to several questions, e.g.:
      a. What is your happiest childhood memory? Please describe it briefly.
      b. What is an interesting fact about you that would surprise other people?
      c. What was your favorite class in high school or college, and why?
      d. Who is your role model, and what do you admire about them?
      e. What is your favorite movie? What did you like about it?
   2. Experiment 1:
      a. Real groups of 3 had mean accuracy of 62% (61% truth; 64% lies)
      b. Individuals had mean accuracy of 53.6% (57% truth; 51% lies)
      c. Simulated groups of 15 had mean accuracy of 57% (65% truth; 51% lies)
   3. Experiment 2:
      a. Real groups of 3 had mean accuracy of 60% (64% truth; 57% lies)
      b. Individuals had mean accuracy of 54% (61% truth; 56% lies)
      c. Simulated groups of 15 had mean accuracy of 49% (75% truth; 47% lies)
   4. Experiment 3: based on high-stakes videos of UK game show contestants
      a. Real groups of 3 had mean accuracy of 53% (52% truth; 54% lies)
      b. Individuals had mean accuracy of 48% (53% truth; 44% lies)
      c. Simulated groups of 15 had mean accuracy of 30% (58% truth; 34% lies)
   5. Results indicate that group decision making more accurate because individuals bring up

points that add to the information held by the group, and increasing accuracy (a synergy mechanism).

H. **Culhane et al (2015)**. Examined whether true/lie judgments better in dyads than by individuals. Also examined whether type of lie (mock transgressions versus real transgression [RT]) influenced truth-telling ability, and whether audio vs auidio/video presentation mode influenced deception detection accuracy.
1. 282 university students evaluated eight recorded statements for truthfulness. True and false statements were elicited through cheating paradigm (i.e. RTs) or mock transgression paradigm. Ps either viewed and listened to statements, or only listened to audio recording.
2. Results:
   a. Dyads no more accurate than individuals
   b. Ps more accurate in detecting deception in RT situations
   c. Highest accuracy rates in RT audio condition.

I. Wu et al (2015). In a study conducted in China, authors manipulated the motivation level of truth/lie judges by varying monetary reward for accurate judgments. Audio-only descriptions of travel, half true, half false made by participants rewarded financially for being judged truthful.
1. Overall accuracy rate = 46.9%,; lie-accuracy = 37.5%; truth-accuracy rate = 56.30% (above chance)
2. Participants had truth bias, judging 59.4% of all statements as true
3. No male-female differences in accuracy
4. Truth-accuracy (56.30%) higher than lie-accuracy rate (37.53%).
5. Motivation increased truth-accuracy, and total accuracy, but not lie accuracy, controlling for the higher truth bias of highly motivated participants.

J. **Street & Richardson (2015).** Demonstrated that people general have a truth bias, i.e., they expect that the statements of others are or will be true. Changing that expectation reduces or eliminates the truth bias early in a series of true/false judgments, but as time goes on, the truth bias returns.

IV. **Lie and/or Deception Detection Methods**.

A. Introduction:
1. What about getting help in detecting deception from devices that measure physiological changes that take place during a lie?
2. All commonly used -lie detection methods do not detect lies, but detect the negative emotions - guilt especially - that accompany telling a lie. The most common such method involves the use of the lie detector, or polygraph, to detect anxiety.

B. Distinction between 'lie detector' and 'polygraph':

1. A lie detector is any person, process, or device that is designed to detect deception.
2. The polygraph is a device that simultaneously records a number of physiological (or other) responses. It can be used for a number of things besides lie detection.

C. The polygraph test measure the physiological concomitants of experienced emotions, usually some subset of the following
   1. Blood pressure: increases with anxiety
   2. Heart rate: increases with anxiety
   3. Respiration:
   4. Skin conductance, or skin resistance (galvanic skin response, GSR): Are inverses of each other, but since conductance measures are more stable, are preferred, especially in research.): Increases with anxiety.
   5. Body temperature:

D. Several ways to conduct a polygraph test, but all compare Ss physiological responses to critical test questions with their responses to a set of baseline or control questions:

E. **Relevant/Irrelevant test**: Developed in 1917 by William Marston; refined in 1921 by John Larson. Involves two types of questions:
   1. Relevant questions about the crime or event in question
   2. Irrelevant questions about nothing in particular
   3. Responses to both types compared, but responses almost always higher to Relevant than to Irrelevant questions, making interpretation problematic.
   4. Not used much in legal cases as a result

F. **Control Question Test** (CQT): Sometimes called the 'Comparison' Question Test. Most common procedure in criminal investigations and elsewhere. Involves 3 types of questions:
   1. **Relevant** questions about the crime or event in question
   2. **Irrelevant** questions about the individual (age, residence, etc.)
   3. **Control questions** about individual's history of honesty. Designed to arouse anxiety in most subjects, and to produce lies:
      a. "Have you ever stolen money?"
      b. "Have you ever driven while intoxicated?"
      c. "Did you ever do anything illegal before the age of 25?"
      d. "Have you ever tried to hurt someone to get revenge."
   4. Total of 10 questions, including 3 Relevant/Control pairs.
   5. Responses to irrelevant questions are not scored, but responses to Control and Relevant questions are compared:
      a. Guilty Ss assumed to respond more to Relevant than Control questions
      b. Innocent Ss assumed to respond more to Control than to Relevant questions
   6. Presumed more sensitive and accurate than R/I test, because directionality of comparison between Control and Relevant questions opposite in guilt and innocence.
   7. Procedure involves Pre-test, Test, and Post-Test phases:
      a. **Pre-Test Phase**:
         (1) Ca. 60 mins.
         (2) Learns personal details of subject, and develops Irrelevant questions.

(3) Tries to convince individual of test accuracy of the test, often by rigging task. E.g., 'stim test'; subject asked to choose playing card and place it back in deck. Examiner shuffles cards, shows them individually to subject who says 'no' to each card. Examiner then picks card subject chose.
b. **Test Phase**: Questions are asked and responses to Relevant/Control pairs are scored using 3-level Control - Relevant measure:
(1) Scored +1, +2, or +3 if Control > Relevant
(2) Scored -1, -2, or -3 if Relevant > Control
(3) Score of zero if both responses the same
c. **Post-Test Phase**: Interpreting results by summing scores over all Relevant/Control pairs and over all physiological measures. 3 measures and 3 pairs would yield 9 scores to sum. If scores sufficiently large and negative, deception is assumed. Scores close to zero are inconclusive.
d. Variations on the Control Question Test
(1) **Directed Lie Test (DLT)**: Very broad question used for control questions, and subject told to lie to them all. E.g. "Have you ever told a lie?"
(2) **Positive Control Test (PCT)**: Relevant question, asked twice, used as control. Subject answers truthfully on one repetition, falsely on the other.
8. **Criticisms** of the Control Question Test and its derivatives:
a. Innocent individuals might respond more to Relevant than to Control questions because of anxiety over the possibility of being convicted of the crime involved.
b. Guilty Ss may not react strongly to Relevant questions because they have been asked so many times before.

G. **Guilty Knowledge Test (GKT)**: Developed in late 1950s by David.T. Lykken.
1. Does not detect lies, but determines whether subject has knowledge of crime or event that only the perpetrator would know.
2. Usually involves just GSR response; based on fact that people react more strongly to important or novel information than to unimportant or familiar information.
3. Series of 5-alternative M-C questions about crime. Correct answer something only person involved in crime would know.
4. Someone with guilty knowledge will react more strongly when hearing that alternative than to any others. Probability of doing so by chance is 1 in 5, so can work out chances of multiple reactions to correct alternative.
5. Not routinely used in North America, but used in other countries (e.g., Israel, Japan)
6. Pros and cons
a. Cannot be easily used unless there are a number of aspects of the crime that are unknown to the public
b. Does not require the subject to respond to any questions at all

H. Is polygraph a reliable and valid method of detecting deception?
1. Hard to determine the accuracy of polygraph tests, since determining accuracy requires that we know the truth. Several ways of assessing lie detection:
   a. **Mock crime studies**: Ss commit mock crime (or not), and are then questioned about it. While researcher then knows truth, level of anxiety about lying will probably be lower than in real world. Should lead to underestimate of polygraph validity.
      (1) 'Guilty' subjects have little or no incentive to try to beat the test, as they would in real life.
      (2) 'Innocent' subjects unlikely to be concerned about relevant questions.
      (3) Low in **'ecological validity'**
   b. **Field studies**: Compare the scoring of the original examiners with that of 'blind' examiners. Problem is knowing what the truth is. Even if both examiners agree, results could still be faulty.
      (1) Even if suspect later convicted, we cannot be sure that he or she was guilty.
      (2) Most studies used confessions to classify subjects as innocent or guilty - a problem: Confession may be false, or may be made because of failed polygraph test, so not independent measure of truth. Would inflate estimate of polygraph accuracy.
      (3) Guilty Ss who pass the polygraph test have no incentive to confess their guilt
   c. **Field-Analogue studies**:
      (1) Ss given opportunity to commit 'crime' (e.g., cheat on test) in a way that is detectable to researchers. Then given polygraph test.
      (2) Preserves anxiety of Ss and real-life analogy
      (3) Ethical questions raised about the deception involved

I. Studies Assessing the Validity of the Polygraph using Field, and Field-Analogue Studies
1. To assess accuracy of binary (e.g., right/wrong; innocent/guilty) detection methods, need four pieces of information:
   a. **Hits**: A guilty subject is correctly identified as guilty.
   b. **Misses**: A guilty subject is wrongly classified as innocent.
   c. **False alarms**: An innocent subject is wrongly classified as guilty.
   d. **Correct Rejections**: An innocent subject is correctly classified as innocent.
   e. For CQT or GKT, there is an additional category - **Inconclusive**
2. **Honts & Raskin (1988)**: Field study of the **Directed Lie Test**. Biased because 'actual' guilt based on whether subject confessed or not. Note:
   a. High level of detection accuracy for Guilty lie-tellers (Hits); low for False Alarms (innocent called Guilty)
   b. 10% inconclusive
   c. Big drop in Correct Rejections (Innocent called innocent) when reviewer is blind. Suggests that some other cues than the polygraph record are being used - examiner spends time with subject in Phase I, and is present to see non-physiological cues to deception.

	'Actually' Innocent	'Actually' Guilty
Judged Innocent	91%/62%	8%/8%
Judged Guilty	0%/15%	92%/92%
Inconclusive	9%/23%	0%/0%

* **Blind examiners ratings under slash**

3. **Patrick & Iacono** (1991): RCMP field study of CQT involving 400+ cases. Estimated 'truth' from case files (e.g., confessions by subjects who did not take a polygraph test, , etc.). had trouble finding independent verification for judged-guilty subjects. (check 2006 also)

	'Actually' Innocent	'Actually' Guilty
Judged Innocent	73% / 30%	0% / 2%
Judged Guilty	8% / 24%	98% / 92%
Inconclusive	19% / 46%	2% / 6%

4. Summary of several CQT reviews: Shows proportion of 'innocent' and 'guilty' subjects detected in a number of different studies. (**Vranhag & Vrij, 2005**) - See Brewer & Williams, p. 56

Review	Innocent	Guilty
Ben-Shakhar & Furedy (1990); N=9	72%	84%
Carroll (1991); N=3	53%	83%
Honts & Perry (1992); N=3	59%	86%
Iacono & Patrick (1997); N=3	56%	84%
Saxe et al (1985) OTA Report; N=10	78%	88%
Lykken (1998); N=4	61%	86%
Raskin & Honts (2002); N=4	59%	89%

5. **Ginton et al (1982)**:
   a. Subjects were 21 police officers attending a course
   b. Given an aptitude test, with opportunity to cheat in a detectable way.
   c. 7 of 21 officers cheated, and all 21 were called back a few days later to take lie detector test (2 cheaters and one non-cheating officer declined, and 3 cheaters confessed beforehand)
   d. 13 non-cheaters and 2 cheaters given CQT, and results rated by blind evaluators
      (1) only 1 of 2 cheaters and 7 of 13 non-cheaters were correctly identified

J. Generally, CQT test OK at detecting lies, but much poorer job identifying truth - at least for blind evaluation. Rates better for involved evaluators, indicating that they are using cues other than CQT results to make their assessments.
   1. Subjects may show arousal when control questions are arousing: E.g., innocent man, accused of murdering wife, is asked questions about his wife.
   2. Subject may fear result of question and be aroused even though answering truthfully.

K. Guilty Knowledge Test (GKT) appears to have the opposite bias
   1. **Ben-Shakhar & Elaad (2003)**: Found 95% of innocent subjects correctly detected, but only 85% of guilty subjects. Outcomes better when studies:
      a. Are mock crime studies
      b. Provide motives to succeed
      c. Allow verbal responses to alternatives
      d. At least 5 questions
   2. **Elaad (1990)**, a field study, found:
      a. 98% of innocent subjects correctly identified

    b. Only 42% of guilty subject identified
  3. **Elaad, Ginton & Jungman (1992):** Also a field study, found:
    a. 94% of innocent detected correctly
    b. Only 76% of guilty subjects correctly detected

V. Fooling the polygraph

  A. **Honts, Raskin, & Kircher (1994)**: Subjects could escape detection by polygraph in mock crime study with 30 minutes of instruction on the rationale underlying CQT. Were able to beat polygraph by counting backwards by 7 from number >200, or biting tongue, pressing toes on floor, during control questions.

  B. In lab mock crime studies, psychopaths detected in deception at same rate as non-psychopaths:
   1. Raskin & Hare (1978)
   2. Patrick & Iacono (1989)

  C. **NRC (2003)** report concluded:
   1. Theoretical rationale for polygraph is weak, especially re differential fear, arousal, or other emotional states to relevant and comparison questions.
   2. Physiological responses measured by polygraph not uniquely related to deception.
   3. Lab tests likely to overestimate accuracy in field practice by unknown amount.
   4. Polygraph test accuracy may be degraded by countermeasures, especially incentives are high
   5. Computerized analysis of polygraph records may improve accuracy modestly, but this potential has not yet been demonstrated.
   6. Polygraph exams may elicit admissions and confessions, deter undesired activity, and instill public confidence, apart from their validity.

  D. Admissibility in court
   1. Not admissible in Canada.
   2. First submitted in evidence in U.S. in Frye vs United States (1923) - just two years after the invention of the modern polygraph in 1921. Polygraph evidence (by William Marston) deemed inadmissable because it was not generally accepted in the field.
   3. Recently (1998) rejected by U.S. Supreme Court on the grounds that it would take over the role of the jury in determining the credibility of a witness.

VI. Brain scanning (ERPs; event-related potentials, or evoked-response potentials)

  A. When a stimulus is presented, there is a measurable response to it - an **ERP**

B. Measured in the same way as an EEG. Distinctive waves names for the amount of time they occur after a stimulus is presented, and for their direction (positive or negative)

C. **P300**: Positive-moving potential ca 300 msec after presentation of novel or **significant**, meaningful stimulus, most strongly in the parietal area.
   1. Usually with **Guilty Knowledge Test**; responses to crime-related vs control items compared. For innocent person, crime-related items should be less important. Called **ERP-GKT**.
      a. Used in 2001 as part of overturning murder conviction of Terry Harrington in Iowa. Passed ERP-GKT, and showed lower arousal to alibi-related items than to crime-related items.
      b. Used in Ray Slaughter's (2004) death-row appeal in Oklahoma. Passed Farwell's test.
   2. Developed by Dr. Emanuel Donchin and **Dr. Lawrence Farwell** (originally research associate at Harvard). Farwell founded Brain Fingerprinting Laboratories, Inc to market the technique, which he has trademarked as "Brain Fingerprinting". Using a technique - including P300 - that he calls MERMER (Memory and Encoding Related Multifaceted Electroencephalographic Response): See:
      a. Http://www.brainwavescience.com
      b. http://www.brainwavescience.com/Role%20of%20BF%20in%20Criminal.php
      c. See also critical analysis by Peter Rosenfeld of Northwestern: http://www.srmhp.org/0401/brain-fingerprinting.html
   3. Evaluation of P300-GKT (**Iacono & Patrick, 2006**)
      a. Laboratory investigations suggest equal or higher success rate for P300-GKT as for regular polygraph in detecting those with guilty knowledge. Like polygraph, seems better at detecting guilty than at clearing innocent.
      b. **Rosenfeld et al (2004):** Suggests countermeasures may exist for guilty individuals on irrelevant target presentations to defeat P300-GKT:
         (1) pressing finger against a leg
         (2) wiggling big toe
         (3) imagining being slapped
      c. **Abootalebia, Moradia & Khalilzadeh (2006).** Rates of correct detection in guilty and innocent subjects 74–80%.
      d. **NRC (2003)** review:
         (1) no indication of whether P300-GKT better than traditional polygraph measures
         (2) no indication whether combining polygraph with P300 might yield better results than either alone.
         (3) not known whether simple countermeasures could generate brain responses to comparison questions similar to those to relevant questions.

D. n P300-based GKT, there are typically three kinds of stimuli presented to subjects:
   1. Probes (P), which are related to concealed information and are known only to the guilty person and authorities. In fact, the guilty subject is expected to know these stimuli, but the innocent one is not.
   2. Irrelevants (I), which are items unrelated to the criminal acts and thus unrecognized by

all subjects (guilty or innocent).
3. Targets (T), which are usually irrelevant items, but all subjects are asked to do a task (for example, pressing a button or count increasing) whenever they see the T, but not when they see a P or an I.
4. "The number of irrelevant stimuli is many times greater than the numbers of the other two types; and therefore probes and targets are rare stimuli. The T stimuli force the subject to pay attention to items, because failure in responding to these stimuli suggests that the subject is not cooperating. Also, the T stimuli are rare and task relevant and thus evoke a P300 component that has been used in subsequent analysis of the probes as a typical P300 of the subject [2] L.A. Farwel and E. Donchin, The truth will out: interrogative polygraphy (lie detection) with event-related potentials, Psychophysiology 28 (1991), pp. 531–547.[2], although this assumption that the T-P300 is a classical rendition of standard P300 has been shown to be sometimes wrong [16]."
5. "There are conventionally two approaches in the analysis of signals and detection of deception in P300-based GKT. In the first – used by Rosenfeld et al. – the amplitude of P300 response in P and I items are compared [16]. In guilty subjects, one expects P > I while in innocents P is another I and so no P–I difference is expected. Based on this theory, the Bootstrapped amplitude difference (BAD) method has been introduced and used by Rosenfeld."
6. "The second approach, introduced by Farwell and Donchin [2], is based on the expectation that in a guilty person, the P and T stimuli should evoke similar P300 responses, whereas in an innocent subject, P responses will look more like I responses. Thus, in this method we called here Bootstrapped correlation difference (BCD), the cross correlation of P and T waveforms is compared with that of P and I. In guilty subjects, the P–T correlation is expected to exceed the P–I correlation and the opposite is expected in innocents."

E. **N400**: A negative-moving potential that occurs about 400 msec after the presentation of a novel stimulus. The N400 reliably occurs in response to semantic incongruity, e.g., when a sentence ends in an unexpected way. Still in the early stages of investigation.

F. **fMRI**: functional magnetic resonance imagery. Measures change in blood flow using changes in magnetic field around head. Several studies demonstrate differences in regions activated during deception, but technique not so far been applied to detecting lies in forensic setting, partly because of time (2-3 hours) and expense in using fMRI procedure.
1. **Spence et al (2001)**:
2. **Langleben et al (2002)**:
3. **Ganis et al (2003)**:
4. **NRC (2003)** review: fMRI not presently useful for detection of deception in applied settings, and complexity of analysis may be prohibitive for all applications for some time.

G. Thermal imaging of face: Uses high speed camera to record rapid changes in facial blood flow.
1. **Pavlidis, Eberhardt & Levine (2002)**: Individuals committed mock crime (stab mannikin and rob it of $20), or were in control condition. Detected 6/8 guilty subjects

and 10/11 innocent ones using this technique, though technique was poorly described in the paper, and 10 subjects were inexplicably removed from the study.
2. **NRC (2003)** review comments on the above study:
   a. Uses only subset of examinees; no info on selection process. Gives no info re decision criteria used to judge deceptiveness from thermographic data
   b. A flawed and incomplete evaluation based on small sample
   c. No cross-validation of measurements, no blind evaluation.
   d. Does not provide acceptable scientific evidence to support use of facial thermography to detect deception.
3. Has advantage that it can be used without the subjects knowledge.
4. Not widely tested for accuracy; will it will detect innocent at a high enough level?

References

Ben-Shakhar, G. & Elaad, E. The validity of psychophysiological detection of information with the guilty knowledge test: A meta-analytic review. *Journal of Applied Psychology*, 2003, 88, 131-151.

Carter, N.L. & Weber, J.M. (2010). Not Pollyannas: Higher Generalized Trust Predicts Lie Detection Ability. *Social Psychological and Personality Science*, 1 (3): 274

Culhane, S.E., Kehn, A., Hatz, J. & Hildebrand, M.M. (2015). Are Two Heads Better than One? Assessing the Influence of Collaborative Judgements and Presentation Mode on Deception Detection for Real and Mock Transgressions. *Journal of Investigative Psychology and Offender Profiling, 12, 158–170.*

DePaulo, B.M., Lindsay, J.J., Malone, B.E., Muhlenbruck, L., Charlton, K., & Cooper, H. (2003). Cues to Deception. *Psychological Bulletin*, 129, 74-118.

Ekman, P., O'Sullivan, M. & Frank, M.G. A few can catch a liar. *Psychological Science*, 1999, 10, 263-266.

Elaad, E. Detection of guilty knowledge in real-life criminal applications. *Journal of Applied Psychology*, 1990, 75, 521-529.

Elaad, E., Ginton, A, & Jungman, N. Detection measures in real-life criminal guilty knowledge tests. *Journal of Applied Psychology*, 1992, 77, 757-767.

Farwell, L.A. & Donchin, E. The 'brain detector': P300 in the detection of deception. *Psychophysiology*, 1986, 24, 434.

Farwell, L.A. & Donchin, E. The truth will out: Interrogative polygraphy ("lie detection") with event-related brain potentials. *Psychophysiology*, 1991, 28, 531-547.

Farwell, L. A. and Smith, S. S. Using Brain MERMER Testing to Detect Concealed Knowledge Despite Efforts to Conceal. *Journal of Forensic Sciences*, 2001, 46(1), pp. 1-9.

Fiedler, Klaus, Schmid, Jeannette & Stahl, Teresa. What Is the Current Truth About Polygraph Lie Detection? *Basic and Applied Social Psychology*, 2002, Vol. 24, No. 4, Pages 313-324.

Ginton, A., Daie, N., Elaad, E., & Ben-Shakhar, G. A method for evaluating the use of the polygraph in a real-life situation. *Journal of Applied Psychology*, 1982, 67, pp. 131-147.

Granhag, Par Anders & Vrij, Aldert. Deception Detection. In Brewer, N. & Williams, Kipling D. (Eds.) *Psychology and Law: An Empirical Perspective*. (Pp. 43-92). New York: Guilfored Press, 2005.

Honts, C. R., Kircher, J. C., & Raskin, D. C. (1995). Polygrapher's dilemma or psychologist's chimaera: A reply to Furedy's logico-ethical considerations for psychophysiological practitioners and researchers. *International Journal of Psychophysiology, 20,* 199-207

Honts, C.R. & Raskin, D.C. A field study of the validity of the directed lie control question. *Journal of Police Science and Administration*. 1988, 16, 56-61.

Honts, C.R., Raskin, D.C. & Kircher, J.C. Mental and physical countermeasures reduce the accuracy of polygraph tests. *Journal of Applied Social Psychology*, 1994, 79, 252-259.

Iacono, William G. & Patrick, Christopher J. *Polygraph ("Lie Detector") Testing: Current Status and Energing Trends.* In I.B. Weiner & A.K. Hess, *The Handbook of Forensic Psychology*, 3rd Edition. John Wiley & Sons, 2006, pp. Pp. 552-588.

Klein, N. & Epley, N. (2015). Group discussion improves lie detection. *Proceedings of the National Academy of Sciences*, 112(24). Pp. 7460-7465.

Levine, T.R. & Bond, C.F. (2014). Direct and Indirect Measures of Lie Detection Tell the Same Story: A Reply to ten Brinke, Stimson, and Carney (2014). *Psychological Science* OnlineFirst, published on June 25, 2014.

Lykken, D.T. (1959). The GSR in the detection of guilt. *Journal of Applied Psychology*, 43, 385-388.

Lykken, D.T. (1960). The validity of the guilty knowledge technique: The effects of faking. *Journal of Applied Psychology*, 44, 258-262.

Lykken, D. T. (1974). Psychology and the lie detection industry. *American Psychologist*, 29, 725-739.

Lykken, D.T. (1978). Uses and abuses of the polygraph. In: H.L. Pick (Ed.). Psychology: From research to practice. New York: Plenum Press.

Lykken, D.T. (1998). A Tremor in the Blood: Uses and Abuses of the Lie Detector. New York: Plenum Trade.

O'Sullivan, M. (2003). The fundamental attribution error in detecting deception: The boy-who-cried-wolf effect. *Personality and Social Psychology Bulletin*, 29, 1316–1327.

O'Sullivan, M. (2005). Emotional intelligence and deception detection: Why most people can't 'read' others, but a few can. In R. E. Riggio & R. S. Feldman (Eds.), *Applications of Nonverbal communication* (pp. 215–253). Mahwah, NJ: Erlbaum.

O'Sullivan, M., & Ekman, P. (2004). The wizards of deception detection. In P. A. Granhag & L. A. Stromwall (Eds.), *Deception Detection in Forensic Contexts* (pp. 269–286). Cambridge, UK: Cambridge Press.

O'Sullivan, M., Ekman, P., & Friesen, W. V. (1988). The effect of comparisons on detecting deceit. *Journal of Nonverbal Behavior*. 12, 203–216.

Patrick, C.J. & Iacono, W.G. Validity of the control question polygraph test: The problem of sampling bias. *Journal of Applied Psychology*, 1991, 76, 229-238.

Pavlidis, I, Eberhardt, N.L., & Levine, J.A. Seeing through the face of deception. *Nature*, 2002, 415, 35.

Porter, S., Woodworth, M., & Birt, A. Truth, lies, and videotape: An investigation of the ability of federal officers to detect deception. *Law and Human Behavior*. 2000, 24, 643-658.

Privette, G. (1983). Peak experience, peak performance, and flow: A comparative analysis of positive human experiences. *Journal of Personality and Social Psychology*, 45, 1361-1368.

Smith, N. (2001). Reading Between the Lines: An Evaluation of the Scientific Content Analysis Technique (SCAN). Police Research Series Paper 135. London: Home Office Policing and Reducing Crime Unit.

Street, C.N.H & Richardson D.C. (2015). Lies, Damn Lies, and Expectations: How Base Rates Inform Lie–Truth Judgments. *Applied Cognitive Psychology*, 29: 149–155.

Ten Brinke, L., Stimson, D. & Carney, D.R. (2014). Some Evidence for Unconscious Lie Detection. *Psychological Science*, Online, March, 2014

Vranhag, P.A. & Vrij, A. Deception Detection. In N. Brewer & K.D. Williams, *Psychology and Law*, Guilford Press, 2005. Pp. 43-92.

Vrij, A. Nonverbal communication and credibility. In Memon, A., Vrij, A., & Bull, R. (Eds.) *Psychology and Law: Truthfulness, Accuracy, and Credibility*. (Pp. 32-58). London: McGraw-Hill, 1998.

Wolpe, P.R., Foster, K.R. & Langleben, D.D. (2005). Emerging Neurotechnologies for Lie-Detection: Promises and Perils. *The American Journal of Bioethics*, 5(2), 39-49.

Wu, S., Cai, W. & Jin, S. (2015). Motivation Enhances the Ability to Detect Truth from Deception in Audio-only Messages. *Journal of Investigative Psychology and Offender Profiling*, 12: 119–126.

Eyewitness Testimony

I. Introduction to eyewitness testimony research

　A. For as long as there have been courts and trials the gold standard of evidence has been the testimony of someone who saw the events involved in the trial. Because we are intensely visual animals, we rely on visual information more than that from other senses.
　　1. "Seeing is believing"
　　2. "I'm from Missouri - me you've got to show."
　　3. "I'll believe it when I see it with my own eyes."
　　4. Indeed, "I see" is often used as a synonym for understand.

　B. In dozens of movies and television programs, a standard dramatic scene is one in which the law enforcement officers confront a suspect and get him to confess with the damning statement that "we have a witness who places you at the scene of the crime."

　C. But consider the fallowing situations:
　　1. Arthur Lee Whitfield released from prison in 2003 after serving 22 years of 63-year rape sentence. Both victims identified Whitfield at trial as definitely the man who raped them. Freed by exonerating DNA evidence.
　　2. In early 2005, Minnesota police officer David Hansen arrested and charged with rape and kidnapping after being identified by the victim from 1,500 photos. Cleared by DNA evidence and charges dropped {http://www.truthinjustice.org/witness.htm}
　　3. On November 24, 2006, CNN reported a review of the 1992 conviction of Marlon Pendleton, convicted for rape on the basis of the victim's identification. A DNA test cleared him.
　　4. James Lee Woodard freed in April 2008 after spending 27 years in a Texas prison after being convicted of murdering his ex-girlfriend in 1980. He was freed by DNA evidence after having been convicted on the basis of testimony from two eyewitnesses. Mr. Woodard, now 55, was imprisoned longer than any other inmate freed by DNA evidence. District Judge Mark Stoltz, in a brief hearing before Woodard's release, told Woodard "No words can express what a tragic story yours is."
　　5. Dr. Daniel Thompson, Australian psychologist & memory expert. Unexpectedly detained by the police on suspicion of rape since he matched perfectly victim's description of rapist. Fortunately, his alibi couldn't have been better: Just before the crime occurred, he was on a live television program describing how one could improve one's memory for faces!

　D. Of the hundreds of convicts who have eventually been exonerated of rape or murder, it is estimated that 90% were convicted totally or in part because of eyewitness testimony.

　E. Now we realize that eyewitness testimony must be taken with a grain of salt.
　　1. Not that this is a new realization: James McKeen Cattel's 1895 research on the inaccuracy of Columbia University students memory and William Stern's 1901 research were about the inaccuracy of eyewitness memory. Apart from the work of Stern and

others late in the 19th century, modern research began in 1970s with Elizabeth Loftus.
2. It became more significant, influential & controversial - in 1980s and especially 1990s when issue of possibly false recovered memories of childhood sexual abuse was raised re high profile cases in the U.S., and a few later in Canada.

II. The Memory System

A. In eyewitness testimony, a witness is not reporting what he or she saw, but what he or she remembers. That memory is certainly based on what was seen, but also influenced by many other things as well.

B. Function of memory system is our survival. To do that:
1. **We don't have to remember everything**: We don't store everything that happens to us because we don't have to.
    a. We typically don't recall much of anything from the first two years of life.
    b. We don't store much info about repetitive events.
        (1) **Blake, Nazarian & Castel (2015)** tested P's ability to draw or recognize the Apple logo. Only 1 in 85 particpants could draw the essential elements of the logo, and less than half of participants recognized the logo from an array of 8 possibilities.
    c. We feel we remember more because we combine bits of sensory information into complete memory. What goes into our **reconstruction** of the past is not just things that happened, but things that probably happened, and things we wanted to happen, and things we expected to happen - even if they did not.
2. In order for memory to perform its function, **memory doesn't have to be 100% accurate all the time**.

III. Three interrelated processes in memory formation and retrieval:

A. **Encoding**: To 'see' something, we must pay attention to it. We don't pay attention to everything so many things not 'seen'.
1. **Exposure duration: Shapiro & Penrod (1986)** did meta-analysis of eight eyewitness lineup studies. Found shorter exposure durations led to lower correct identification, and higher false identification rates.
2. **Arousal level**: How well we pay attention and how well er encode also depends on levels of arousal. Moderate levels of arousal best. Both low and high arousal reduce encoding accuracy.
3. **Distraction**: Reduces attention to information:
    a. **Inattentional blindness: Simmons & Chabris (1999)**: Ss see video of people passing basketball back and forth; asked to count passes made by one person.

Woman in gorilla suit crosses screen, thumps chest, moves on. About 10 seconds on screen. About half Ss did not see her.
http://viscog.beckman.uiuc.edu/grafs/demos/15.html

    b. **Change blindness**: We may not notice when things change. Researcher stops person in street to ask directions. While directions given, workers carrying door between E and subject, and accomplice replaces E. Subject usually continue talking to E, often doesn't notice Experimenter has been replaced. (**Simons & Levin, 1998**)

    c. Increased number of perpetrators (**Clifford & Hollin, 1981**) degrades memory.

    d. **Weapon focus**: When weapon present in crime event, witnesses focus on it, can recall aspects of weapon and hand that held it in detail, but less attention to other aspects of scene, including offender's face. (**Loftus, Loftus, & Messo, 1987**; **Kramer, Buckhout, & Eugenio, 1990**; **Steblay, 1992**)

4. **Hope1 et al (2012)**: Found that individuals (law enforcement personnel) who had previously performed a physically demanding task had poorer recall and recognition performance compared with control. They provided less accurate info re critical and incidental target individuals encountered during the scenario, recalled less briefing information, and provided fewer briefing updates than control participants did. Exertion also associated with reduced accuracy in identifying critical target from lineup.

5. **Distinctiveness**: Distinctiveness improves recall, but distinctive 'flashbulb' memories not as accurate as once believed. May be more so for events that have high personal relevance.

    a. Ulric **Neisser (1986)** asked first-year university students how they first heard about Challenger shuttle explosion, then asked again two and a half years later. None of later accounts matched original report. Neisser says about 30% of recollections 'wildly inaccurate'.

        (1) E.g., description by one student in January 1986: "*I was in my religion class and some people walked in and started talking about the [explosion]. I didn't know any details except that it had exploded and the schoolteacher's students had all been watching, which I thought was so sad. Then after class I went to my room and watched the TV program talking about it and I got all the details from that.*"

        (2) Same student in September 1988: "*When I first heard about the explosion I was sitting in my freshman dorm room with my roommate and we were watching TV. It came on a news flash and were we both totally shocked. I was really upset and went upstairs to talk to a friend of mine and then I called my parents.*"

    b. See also: **Neisser & Harsch (1992)**; **Schmolck, Buffalo, & Squire (2000)**; **Talarico & Rubin (2003)**

B. **Storage and Retrieval**: What put in memory is not literal copy of event, but **interpretation** of what happened based not only on what seen, but how we interpret what seen.

    1. **Labelling**: 1932 study asked Ss to remember labelled drawings like these. Some Ss saw drawings labelled 'sun', 'curtains', 'stirrup'. Other Ss saw same pictures labelled 'wheel', 'diamond', 'bottle'. When Ss later reproduced images, their drawings influenced by label given.

2. **Prejudices and biases**:
    a. Buckhout (1974): Ss describe series of pictures seen earlier. One showed dark-skinned man and shorter light-skinned man facing each other on bus. Lighter man dressed as workman, carried what might have been painter's trowel. Dark-skinned man dressed in suit, carried briefcase. When mostly white Ss asked to describe picture, many recalled light-skinned man as taller, or that he who was wearing suit. These distortions reflect in racial prejudices of subjects, and perhaps also their expectations. (This was originally done by Gordon Allport and Leo Postman in a 1947 study of rumour transmission. They showed the picture to one person, who described it to another, who described it to a third, etc. after six retellings, the razor blade in the white man's hand usually shifted to the Black man's hand.)
    b. **Boon & Davies (1996)**: Fans recall of infractions at football game biased by fan's team loyalties. After **Hastorf & Cantril (1954)**.
        (1) Showed Ss film of rough 1951 football game between Dartmouth and Princeton. Many penalties and injuries; had spawned editorials in campus newspapers. All-American Princeton quarterback left game in 2nd quarter with broken nose and concussion. Dartmouth quarterback's leg broken in 3rd quarter. One week after game, Dartmouth and Princeton psych students filled out questionnaire, and authors analyzed answers for those who had seen the game or movie of the game. Two other groups viewed film of game and tabulated infractions for each side. Dartmouth and Princeton students gave discrepant responses. 36% of Dartmouth students and 86% of Princeton students said Dartmouth started rough play; it; 53% of the Dartmouth students and 11% of the Princeton students said that both started it. When shown film of game later, Princeton students saw Dartmouth make over twice as many infractions as seen by Dartmouth students.
    c. See **Bahrick et al (1996)** Ohio Wesleyan University students in first and second years recalled high school grades for research credit. Recalled more 'A' than 'B' grades, more 'B' then 'C' etc.
3. **Inferences**: **Bower and Treyens (1981)**. Ss waiting to participate in experiment in experimenter's office for 30 seconds before taken to another room. Asked about room they had been in. All Ss agreed there was desk and chair, almost a third remembered books when there were none
4. **Interpolated testing/retelling**: Enhanced recall quantity if recall for that material tested in the interim.
    a. **Hyman & Pentland (1996)**: Recall of both true and false childhood events increases with repeated interviewing.
    b. Retrieval-induced forgetting (RIF): Memory for rehearsed items increases, memory for related but unrehearsed items actually decreases. (See **McLeod, 2002; Shaw, Bjork & Handal, 1995**). But false recall may also be better preserved by earlier testing, see **Bergman & Roediger (1999)**.
5. **Leading questions**: Courts know that leading questions can bias testimony, so they are not allowed. You cannot ask questions that suggest a particular answer. You can't ask 'Did you see a gun in the suspect's hand?', but you must ask 'What, if anything, did the subject have in his hand?'
    a. **Harris (1973)**:

- (1) "How tall was the basketball player?" = 79 inches; "How short was the basketball player?" = 69 inches
- (2) "How long was the movie?" = 130 min; "How short was the movie?" = 100 min.
  - b. **Loftus (1975)**: People interviewed about headaches and headache product use:
    - (1) How many other products tried?:
      - (a) "In terms of the total number of products, how many other products have you tried? 1? 2? 3?" = 3.3
      - (b) "In terms of the total number of products, how many other products have you tried? 1? 5? 10?" = 5.2
    - (2) How often do they get headaches?
      - (a) "Do you get headaches frequently, and, if so, how often?" = 2.2/wk
      - (b) "Do you get headaches occasionally, and, if so, how often?" = .7/wk
  - c. **Loftus & Palmer (1974)**: Ss see brief film of low-speed accident involving two cars. After seeing film, Ss estimate speed of cars using one of four different questions:
    - (a) "How fast were the cars going when they smashed";
    - (b) "How fast were the cars going when they bumped?"
    - (c) "How fast were the cars going when they hit?"
    - (d) "How fast were the cars going when they collided?"
    - (2) Speed estimates increased with the violence implied by the word used: Subjects asked using the word 'smashed' gave higher speed estimates [10.46 mph] than subjects asked using the word 'hit' [8.0 mph].
    - (3) A week later, subjects all asked: "Was there glass at the scene of the accident?" Subjects who had been questioned with 'smashed' twice as likely to say yes [16% yes] than subjects questioned using 'hit' [7%]. And there was no glass.
    - (4) Probability of saying yes increased with subjects' earlier speed estimates, regardless of word used in question. In fact, no glass at the accident scene.
  - d. **Loftus & Zanni (1975)**: Participants see film of motor accident, then asked "Did you see a broken headlight?" or "Did you see the broken headlight?" Same results as 'a table/the table' study.
6. **Misleading post-event information**: See **Loftus (1979)** for review. Last two studies above go to the edge - beyond - of leading questions, since they actively suggest something about the event that is not so. Now we are looking at the effect of misleading post-even information (PEI) on subsequent recall.
  - a. **Loftus (1975)** originally supposed that the original information was replaced by the misinformation. Recent studies suggest that it is the accessibility of the correct information that is impaired, not its presence in storage. See **Chandler, 1989, 1991; Christiaansen & Ochalek, 1983**). May be a failure of source attribution or source monitoring.
  - b. **McCloskey & Zaragosa (1985)** argued for two possible reasons for the misinformation effect: (a) deference to the experimenter's (erroneous) account of the event, despite possession of the correct information initially; (b) acceptance of the misinformation due to failure to encode the correct information initially.
  - c. **Source Monitoring**: Keeping track of where a bit of information came from: Did I

see it during the event, or did I learn it afterwards? Errors lead to attributing information to the event that came afterward:
- (1) **Dr. Daniel Thompson**, the Australian psychologist identified as rapist despite being on a live TV program at the time: During sexual assault victim watching that show, and confused Dr. Thompson's face with that of her attacker.
- d. Subjects believe false reports: **Loftus, Donders, Hoffman & Schooler (1989)**: Found that misled Ss responded faster and were more confident than non-misled Ss.

7. Creating completely false memories:
   a. **Loftus & Pickrell (1995)**: 'Lost in the mall' study. Ca 25% of Ss had created full or partial memories after two sessions of questioning.
   b. **Porter, Yuille & Lehman (1999)**: False memories of being attacked by a vicious animal.
   c. Imagining false events increases belief in them. Relevant because several recovered memory procedures - and parts of cognitive interview - involve imagining event:
      - (1) **Garry, Manning, Loftus & Sherman (1996)**: Asked Ss to rate confidence that some childhood events had occurred. Then asked to imagine some events briefly, and new confidence data collected. Found confidence in experiencing imagined events higher than non-imagined events.
      - (2) **Goff & Roediger (1998)**: The more often events are imagined, the more likely Ss are to think they actually happened.
      - (3) **Mazzoni & Memon (2003)**: UK study. Imagining having skin sample taken as a child not only increased confidence that it had happened, but also resulted in the creation of false memories in some Ss. Some memories had lots of detail.

IV. Victim and witness descriptions of offenders:

A. Each year in North America, some psych instructor somewhere is conducting a demonstration of the vagaries of eyewitness descriptions. Confederate comes in, steals something, leaves. Students provide widely discrepant descriptions of 'offender'.

B. We will look at verbal descriptions first (recall), then consider recognition identification from mug books and live or photo lineups.

V. Verbal descriptions of offenders, and their accuracy

A. Two sorts of studies: Mock crime studies, and analyses of descriptions provided by actual crime victims and witnesses. We will consider only archival analyses, most of which (with one notable exception) find that eyewitness descriptions are not very good. What we get also depends on how witness questioned: Did they give free-form recall, or did police ask a series of questions?

B. **Kuehn (1974)**: Analyzed 100 police protocols in injury, rape, and robbery cases in Seattle, WA. Statements taken immediately after event, all perpetrators were strangers.
   1. Witnesses averaged just 7.2 descriptors (maximum of 9, but mostly 8 or 9) - 4 victims unable to provide any details
   2. Most often mentioned (from most to least): Gender; age; height; build; race; weight; complexion, hair color. All features mentioned by more than 70% of victims.
   3. Note that clothing is absent from this list.

C. **Sporer (1992a)**: Analyzed records containing 100 witnesses providing 139 person descriptions of perpetrators of robbery and rape. Half from directly involved, half from witnesses outside crime area.
   1. Number of descriptive details ranged form 1 to 48 (mean of 9.71; SD=7.03)
   2. Nearly 25% of details general: height, age, race, weight, with many referencing some unspoken norm 'average', 'normal', etc.
   3. 31% of descriptions referred to clothes; 30% to face - the majority to the upper half of the face, especially hair.
   4. Hair descriptions least useful, since most easily changed after offense

D. **Lindsay, Martin & Webber (1994)**: Examined descriptions of 105 criminals published in Kingston *Whig Standard*., and compared their completeness with 100 descriptions (of five targets) contained in lab studies.
   1. **Mock crime witnesses**: Most likely to report clothing (99%), hair color (90%), and height (86%), but fewer than 50% reported obvious characteristics - sex, age, race/ethnicity.. Most frequent facial feature described was eyes (43%); all other facial features mentioned less than 25% of time.
   2. **Real crime witnesses**: Most likely to report gender (96%), hair color (38%), and clothing (60%) and race/ethnicity (25%). Descriptions of facial features in fewer than 10% of sample. Mock witnesses provided more details (7.35 features) than real witnesses (3.94 features), and authors conclude: "*The data strongly support our concern that eyewitness descriptions are frequently vague.*" (P. 531)

E. **Van Koppen & Lochun (1997)**: Archival analysis of person descriptions in 431 robberies. 1313 witnesses, 2299 offender descriptions. Descriptions subdivided into 24 permanent aspects of offender (e.g., gender, skin color), and 19 temporary aspects (e.g., clothing, type of mask, etc.)
   1. Completeness of descriptions rather poor: of maximum of 43 possible descriptions, witnesses provided median of 8.
   2. Permanent features mentioned more often (median = 5) than temporary (median = 2);
   3. Fewer than 5% of descriptors referred to inner facial features (e.g., eye color, complexion, nose, mouth, chin, etc.) which are most important in identifying a person.
   4. Majority of temporary descriptions were to clothing and their colors: Hats (51%); hat color (31%); jackets, coats, trousers (28/25/26%); and their colors (28/22/18%, respectively) ** Note the similarity of this to what we hear on Crimestoppers bulletins
   5. Compared witness descriptions with Netherlands police database descriptions:
      a. more descriptions correct than incorrect

b. majority of facial descriptions wrong: % errors = : Eye color: 64%; Nose: 65%; Mouth: 60%; Chin: 62%

F. **Yuille & Cutshall (1986)**: Examined 21 witnesses to single shooting; 13 took follow-up research interview 5 months later. Reported accounts to be elaborate and accurate, even 5 months later. On police interview:
1. 392 action details (82% accurate)
2. 180 person descriptions (76% accurate) - [mean of 9 descriptors]
3. 78 object description details (89% accurate)
    a. 23% errors on height, weight, age (+/- 2 inches, 2 years, or 5 pounds)
    b. 18% errors in style, color of hair; style, color of clothing
    c. almost all descriptions of facial hair (beard and/or mustache) were wrong.

G. Estimates of height and weight: Hard to determine, since most such studies allow for a range of errors in determining accuracy. But correlations between estimated height and weight and actual height and weight are not very high.
1. **Flin & Shepherd (1986)**: Had 588 Ss estimate heights and weights of 14 male targets. Each target accompanied by companion who asked the subject for directions. Companion then returned to ask for estimates of height and weight for both target and subject.
    a. Ss use their own weight and height as 'anchors' in judging weight, height of others, but weight and height of companion did not affect estimates.
    b. Ss tended to underestimate target's height and weight
    c. Also some 'regression to the mean', with overestimates of short/light targets and underestimates of tall/heavy targets

H. Story circulating among U. S. law enforcement agencies: Convenience store robbed and three eyewitnesses work with police to produce facial composites of culprit. Each produces very different face, so police release all three composites on wanted poster. Police later receive message from nearby small town police department: "Have arrested two of the suspects and are hot on the trail of the third."

VI. Police mug books (photo arrays) and identification

A. Effect of mug book viewing on lineup performance: If no overlap between mug book contents and lineup, then no adverse effects. If foil present in mug book is picked as perpetrator, more likely to be selected in lineup as well. (See **Brown, Deffenbacher & Sturgil, 1977** for an early study.)

B. **Memon et al (2002)** found that choosing **ANY** face from photo array increased probability of later choosing an individual in that array from a later live lineup - even if the person chosen from the live lineup was not the person chosen from the photo array.

C. **Goodsell, Gronlund & Neuschatz (2015)** found the usual decrement in lineup accuracy after selection from photo array. Also found that participants asked to chooses SEVERAL photos from the array rather than one, showed no detrimental effect on later lineup identifications. Also found that participants who had made an (incorrect) choice from photo array, and picked same person out of the lineup, failed to recognize the actual perpetrator when present in the lineup.

D. But what are the mechanisms behind the mugshot exposure effect? Two (or three) have been suggested and investigated:
   1. **Familiarity**: Witness at lineup gets sense of familiarity when seeing in lineup a person he saw in the mugshot display, mistaking the familiarity as indicating this is the person at the crime scene. Essentially a case of source confusion.
   2. **Commitment**: Individual has incorrectly chosen a foil from the mugshots, and wants to be consistent in his choice, so chooses same individual when he appears in lineup. Should mean that individual who made no choice from mugshots should be more likely to make no choice from lineup as well. Certainly not to pick someone already seen - but not picked - from mugshots.
   3. Evidence for involvement of both processes.

E. **McAllister, Stewart, & Loveland (2003)**: Looked at size of mug book and position of perpetrator's photo as factors in correct identification. Perpetrator and 69 preceding foils appeared in positions 1-70, 71-140, or 141-210 in static or dynamic computerized mug book.
   1. Found correct IDs lower the later in the book the perpetrator's picture appeared (consistent with previous research)
   2. The later the 69 foils appeared in the book, the lower the number of false positives from among that set.
   3. Probability of correct ID decreased the more pictures were viewed.

VII. Methodological issues in eyewitness identification from photos, lineups, or show-ups

A. Given a table that compares the suspect's status (guilty or innocent) and the witnesses' decision (Guilty of Innocent), we have four decision cells

Judged Status of Individual in Lineup

Actual Status	Guilty	Innocent
Guilty	"Hit" (H)	"Miss" (M)
Innocent	"False alarm" (FA)	"Correct Rejection" (CR

B. To assess accuracy of witness judgments, often use **diagnosticity ratio**, which is rate of Hits over rate of False alarms (H/FA). Sometimes this is calculated as C/(C+FA), which is the proportion of total identifications that are correct IDs. [Later in discussing violence prediction we will come across Positive Predictive Power (PPP), which is H/(H+FA)]
1. The diagnosticity ratio should be high, minimizing chances that person identified as guilty from a lineup is actually innocent.
2. Problems with diagnosticity ratio as measure of accuracy in lineup identification:
    a. An increase in the H/FA ratio can come about in two (or three) different ways:
        (1) an increase in the number of correct identifications ("Hits") without any decrease in the number of incorrect identifications ("False Alarms").
        (2) a decrease in the number of incorrect identifications ("False Alarms") without any increase in the number of correct identifications ("Hits")
        (3) a combination of an increase in Hits and a decrease in False alarms.
3. Sequential, as compared with simultaneous, lineup procedures seem - according to recent research and meta-analyses, to lead to a slight reduction in hits, together with a larger reduction in false alarms, thus increasing the diagnosticity ratio.

C. But why do sequential procedures change the diagnosticity ratio? To make sense of that we have to consider the two processes that are thought to underlie responses to recognition questions: In the identification situation, this would be a question of the form "Is this the person you saw at the scene of the crime?"
1. The first aspect of the recognition process is **discriminability**: To what extent does the person you are looking at in the lineup or photo array resemble the image of the person you saw at the scene of the crime. Let's say, for argument's sake, that we can assign a number between 0 and 100 to that resemblance: No similarity = 0, and perfect identity = 100. This is a measure of the memory ability of the individual, and the value we are really interested in.
2. But there is another process that is involved in a witnesses' selection of a suspect from a lineup, and that is the witnesses' **selection bias**, or **response criterion**. That is, people have to choose a criterion level of discriminability at which to say "Yes, that is the man."
    a. Some witnesses have a very conservative response criterion, requiring (for example) 80% discriminability before they will choose a person as the suspect.
    b. Other witnesses (or even the same witness under different conditions) may have a very liberal response criterion, picking someone as the suspect if the resemblance between him and their memory image is 50% (or even less).

3. Why does this make a difference? Witnesses with the very same discriminability (the same degree of judged similarity between a lineup individual and the memory image of the suspect) may have different response criteria, leading to similar diagnosticity ratios, but for different reasons:
    a. A conservative response criterion leads to both a smaller number of Hits, and a smaller number of False Alarms.
    b. A liberal response criterion leads to both a larger number of Hits and a larger number of false alarms.
4. Consider the example of two students taking a 3CC3 multiple-choice test. Both have the same knowledge, and so the same ability to discriminate the correct answer from the false foils or distractors.
    a. Student A takes test getting 1 point for each correct answer, but losing one point for each incorrect answer, and no points or losses for questions left blank. Student B takes same test, also gets 1 point for each correct answer, but loses 5 points for each incorrect answer, and no points or losses for questions left blank..
    b. You can imagine that second student will be much more cautious in questions she answers, choosing an alternative only if she is **VERY** sure (say 90%) that it is correct, and leaving blank many more questions where she is less than 90% sure.
    c. The first student will be much more likely to answer a question when she is only 50% certain of the answer, and will leave far fewer questions blank.
    d. So what will the difference be between students A and B in terms of their accuracy measured by the diagnosticity ratio? In general, both correct and incorrect answers will be lower for student B (the latter more than the former), so that the C/F ratio will increase.
    e. But since both students know the material equally, well, we will overestimate the ability of student B compared with student A, and assume that discriminability is affected by the conditions of the test, whereas it is really response bias that has been affected, and discriminability remains the same for both students.
5. More recently, measures and techniques from Signal Detection Theory (SDT) have been used to separate discriminability from response bias. In particular, the ROC (Receiver Operating Curve, or Receiver Operating Characteristic) has been used to analyze eyewitness identification data.
    a. The ROC curve plots Hits on Y-axis against False Alarms on x-axis, so ratio of y to x-axis at any point on the curve represents the diagnosticity ratio.
    b. Another measure (or two) can also be calculated from the ROC curve: d' (d-prime), which is a measure of discriminability, separate from response bias.
    c. Another measure that can be calculated from the ROC curve is the area under the curve (AUC), which is also a measure of discriminability.
6. As a result of this separation, ROC analyses are becoming much more common in this research area, and (as we will see later in the course) in assessing estimates of violent reoffending.

D. One area in which the ROC analysis has come into its own is the distinction between simultaneous and sequential lineup procedures, which we will discuss later in this section. Although a number of studies have found the sequential procedure to lead to greater accuracy, as measures by the diagnosticity ratio, an increasing number of studies, using ROC

analyses, find that the difference between simultaneous and sequential procedures is primarily in the response bias, and not in the participant's actual discriminability.
1. **Lindsay & Wells (1985)** argued that participants in a simultaneous lineup used a relative match strategy, choosing the individual who most resembled the perpetrator they remembered. In contrast, participants in a sequential lineup use an absolute strategy, matching their memory to each individual presented in the lineup. The importance of this is that the sequential procedure would lead to greater discriminability; a reduction in incorrect identifications (False Alarms) without a decrease in correct identifications (Hits).
2. **Ebbesen & Flowe (2002)** argued that a sequential lineup led witnesses to adopt a more conservative response criterion, affecting response bias, but not improving discriminability.
3. **Gronlund (2004)** found evidence to support the hypothesis that the simultaneous-sequential lineup difference is most - or totally - due to the adoption of a more conservative criterion in sequential lineups - changing the diagnosticity ratio by reducing response bias, without improving discriminability.
4. **Gronlund et al (2012)** compared showups and simultaneous and sequential lineups using ROC analysis. They found that simultaneous lineups consistently produced more accurate identification evidence than showups, but sequential lineups were sometimes no more accurate than showups, and were never more accurate than simultaneous lineups.
5. **Gronlund, Wixted & Mickes (2014)** note that recent studies comparing simultaneous and sequential lineup procedures using ROC analyses have not supported the notion that sequential lineups are superior to simultaneous ones, and indeed have suggested the opposite. They conclude that "It is not yet clear which lineup procedure will prove to be generally superior, but it is clear that ROC analysis is the only way to make that determination."

VIII. Facial composites from eyewitness descriptions. Three methods, historically:

    A. **Drawings** by police artists:
      1. Used as early as 1910 in UK. More recently used in hunt for Unabomber, Oklahoma bomber, in US.
      2. References: **Bruce, Ness, Hancock, Newman, & Rarity, 2002; Ellis, Davies, & Shephard, 1978; Kovera, Penrod, Pappas, & Thill, 1997; Frowd et al, 2005**
      3. Composites drawn with photograph present often no better than composites constructed from memory (**Heaton-Armstrong et al, 2006**, p.85)
      4. Little research of the accuracy or utility of police drawings. Only 18 full-time artists in the 500 U.S. sheriff's departments as of 2004.

    B. **Mechanical systems** (e.g., *Identi-Kit*; *Photofit*)
      1. **Identikit**: First system developed in 1959 by Hugh MacDonald, a California police officer. Originally consisted of 568 drawings of different facial features (eyes,, lips, noses, etc.) on clear plastic sheets that could be overlain. Anecdotal stories of its

success, but no research.
2. **Photofit**: Developed in UK by Jacques Penry in 1970. Similar to Identikit in function, but used photos of facial features rather than drawings. Some research on accuracy:
   a. **Hadyn Ellis et al (1975)**: Had Ss produce Photofit likeness after viewing photo of one of several white male targets. Panel of judges then attempted to select the photo used from an array of 36. Only 12.5% of judges first choices were correct; 25% if 2nd and 3rd choices included. (This only shows that likeness poor, which could be witnesses fault.)
   b. **Ellis et al (1978a)**: Found no difference in likeness between Photofit images from memory or from a photograph of the target.
   c. Usual racial familiarity effect found: Composites from same-race witnesses more accurate than those from different-race witnesses.
   d. **Christie & Ellis (1981)**: Found verbal descriptions consistently better guide to likeness than Photofit composites. Judges can match descriptions better than composites to photos.
   e. Field tests of Photofit (e.g., **Kitson et al, 1978, Bennett, 1986**) or Identikit (**Levi, 1997**) produce the same poor results.
      (1) **Kitson et al (1997)**: Followed 729 composites made during inquiries by 15 different police forces. Two months after investigations ended, 140 cases solved; in 5% composite entirely responsible for resolution, and 'useful' (33%) or 'very useful' (17%) in solution, but in other cases it was 'not very useful' (20%) or 'no use at all' (25%).
      (2) **Levi (1997)**: Use of Identikit by Israeli police - Only 5 of 54 convictions in 243 cases in which Identikit used were significantly aided by Identikit composite.
C. **Computer systems** (e.g, *Mac-a-Mug*; *E-fit, FACES, Facette*)
   1. **Mac-a-Mug** evaluation:
      a. **Koehn & Fisher (1997)**: Ss meet stranger then make composite of face with Mac-a-Mug Pro which assessed for likeness by judges on 10-point scale. 69% of composites given 1 or 2 rating. Only 4% of composites correctly matched by judges to face in 6-photo array. (Other composites created by skilled operator from life were matched 77% to photos.)
      b. **Kovera, Penrod, Pappas & Thill (1997)**: Ss compile composites of former teachers and classmates. Then given to fellow students who knew the targets, who tried to discriminate them from unfamiliar composites. Judgments of familiarity and confidence made. (And a name, if possible.) But only 3 of 167 names provided were correct.
   2. **E-fit**: Different from most other systems in that it uses photographic-quality features; also marketed with instructions - unlike most others - and training is provided. Involved extensive initial interview, and cued construction of composite based on m-c questions on screen.
      a. **Davies et al (2000)** finds results with Mac-a-Mug and E-fit the same when participants describe faces from memory; E-fit only better when participants work from photographs. Then witness describes same faces, first from memory then with photograph. Judges correctly match 75%+ of composites to photos, but can name only about 35% of composites - even though they know the persons depicted.
      b. **Davies & Oldman (1999)**: Witness assists operator in making composite of one of

four famous faces, both with photo and from memory. Judges only able to name 10% of composites from photo, and 6% made from memory. False naming rate was 25%.
    c. **Brace, Pike & Kemp (2000):** E-fit operator constructs pairs of composites for 48 famous personalities, first from memory and then with aid of photograph.

D. **Computer systems based on genetic algorithms** - still in development
1. Designed to take advantage of Ss ability to discriminate between whole faces rather than to identify individual pieces of a face.
2. See EvoFit: **http://www.psychology.stir.ac.uk/staff/cfrowd/research-index.htm**
    a. http://www.psychology.stir.ac.uk/staff/cfrowd/commercial.htm
    b. **Davies & Valentine (2007)** argue that one problem with current composite production systems is that they are based on a logical rather than a psychological analysis of faces: We don't normally recognize faces by analyzing them into their constituent parts.
    c. 'face-space' - a multidimensional space relating faces by overall similarity, without defining the dimensions along which that similarity exists, might be a better model for face recognition and description. (See **Sirovich & Kirby, 1987**; **Turk & Pentland, 1991**)
    d. Recently, genetic algorithms have been used to search the face-space defined by principle component analysis (PCA) and converge on a desired likeness.
    e. PCA does not represent the texture of hair accurately, which has to be chosen separately from a database.
    f. In GA model, PCA and eigenfaces used to generate a random set of images from the face space. Then S selects closest match or matches, and random changes made around them, new selection, etc. until S cannot choose among equally good choices, or genetic algorithm has failed to converge on desired appearance. Three systems in development:
        (1) Evo-fit (Hancock, Frowd, et al at Stirling)
        (2) Eigen-fit (Solomon et al, University of Kent)
        (3) ID (Tredoux, Rosenthal et al, University of Cape Town)
    g. Right now, genetic algorithm systems performing as well as current composite systems, but hope is held out that they will perform better. Problem in part is hair, which GA systems don't do well, and which is most salient identifier for Caucasian faces (**Ellis, 1986**).

E. Computerized programs: Don't do as well as police artists. Witnesses have a hard time selecting the appropriate pieces.

IX. **Situational variables** that affect lineup ID:

   A. **Changed appearance and disguises**: Offender is disguised at the time of the crime, or changes his appearance between the time of the crime and the attempted identification by the witness. Either reduces accuracy of identification:
   1. See **Shapiro & Penrod (1986)**: Meta-analysis of eight eyewitness lineup studies indicates shorter exposure durations lower correct face identification, and raise false identification rates.
   2. **Read (1995)**: Found reduced identification accuracy when appearance changed with facial hair, hair style, or presence/absence of glasses.
   3. Evidence re disguises at time of the crime is ambiguous - some studies find reduced accuracy, others do not.
   4. Distinctiveness: Distinctive faces are more likely to be attended to, and to be identified correctly. (See **Light, Kayra-Stewart & Hollander, 1979**)

   B. **Cross-race effect.**: We all have more difficulty distinguishing people of different race or ethnic background than we do distinguishing people who look like us. Probably the result of more experience. We are better at discriminations we are experienced in making.
   1. Lower proportion of correct ID from target-present lineups for different race
   2. Higher proportion of false alarms from target-absent lineups for different race
   3. See **Meissner & Brigham (2001)**

   C. **Cross-age, cross-gender** issues: Recent research suggests that ID of opposite sex, and of different aged person, may also be impacted negatively.
   1. **Wright & Sladden (2003)**: Own gender bias.
   2. **Wright & Stroud (2002)**: Own-age bias

X. **Retention variables** that may affect correct recognition: What happens between the event and testing?

   A. **Retention interval**: Recall declines with delay of testing

   B. **Verbal overshadowing Effect (VOE)**: Subjects make fewer correct visual identifications if they have given a verbal description beforehand. (See *Applied Cognitive Psychology*, 2002, Vol. 16(8), pp. 869-997. See also the meta-analysis by **Meissner & Brigham, 2001**)
   1. Although the overshadowing effect is small, it is consistent, and means that witnesses who give a verbal description are 25% more likely to misidentify the offender. But incorrect rejections and false alarms have not been considered, so the effect of verbal overshadowing on them is not known. Not all studies find the effect.
   2. Relevant to providing verbal description to police artist, and to attempting to construct facial composites from pieces, which also involves local processing mode.
   3. Not clear why this happens, and there are three theoretical account of it:
      a. Original hypothesis is that verbal encoding of the description interfered with later

retrieval of original image ('**retrieval-based interference**')
- b. Interestingly, description of any face impairs later retrieval of an offender's face, which does not fit with retrieval-based interference.
- c. **Finger (2002)** showed that engaging in nonverbal tasks (listening to music, completing a maze) eliminated the effect.
- d. **Transfer inappropriate retrieval**? **(Schooler, 2002)**. Asking for verbal description engages verbal processing, and inhibits use of nonverbal processes involved identification.
- e. **Macrae & Lewis (2002)**: Problem is not verbalizing, but the development of a local processing mode as a result. global processing helps identification, but local reduces it. Compared Ss who read whole letters with those who read the letters of which the letters were made.
- f. But see **Clare and Lewandowsky (2004)**:
- g. **Frowd, C. D. & Fields, S. (2011)**: Found that VOE interfered with construction of a traditional facial composite produced by selecting individual facial features.
    - (1) Ps looked at unfamiliar target and two days later constructed composite after either describing the face (verbal no-delay), without describing (no-description) or 30 minutes after describing (verbal delay).
    - (2) Composite quality was poor overall, but worse in verbal no-delay group relative to no-description group.
    - (3) Same overall accuracy in verbal delay and no-description conditions, suggesting a 'release' from overshadowing.
    - (4) Data suggest that witnesses to real crimes should not proceed directly from face description to feature selection.

C. **Memory transference**:
1. Subject may identify as a suspect or offender someone who was seen at the crime (e.g. a bystander), or who has been seen frequently at the scene of the crime (e.g. a bank). Or witness may have been shown a mug shot of the person, and then mistake the sense of familiarity at the lineup to having seen him at the crime scene.
2. Not all studies of this effect find it, but see **Ross et al (1994)**. Apparently, memory confusion so that witness thinks suspect and innocent person are the same person ('memory blending')..
3. unanswered questions about the effect: (**Brewer et al, 2005**, in Brewer & Williams)
    a. How similar do bystander and offender have to be before the effect occurs?
    b. Does time between exposure to bystander and exposure to offender reduce the effect?
    c. Would the effect disappear if witness exposed to lineups with both persons present?
    d. Would using highly similar foils in lineup prevent the effect?
    e. Would sequential lineup prevent the effect?

D. **Context reinstatement**: Is recognition better at the scene of the crime, or in the same emotional state?
1. **Shapiro & Penrod (1986)** found more correct identifications from target-present lineups, but also more false alarms from target-absent lineups. Suggests to Brewer et al

(2005) that context reinstatement simply increases likelihood of making a choice, not overall accuracy.
2. Some studies (**Smith & Vela, 1990**) find imagined reinstatement of context has no effect on accuracy, while others (**Malpass & Devine, 1981**) find an improvement when imagined reinstatement combined with, e.g. detailed event info.

XI. **The identification test** itself:

A. Lineup identification performance is uncorrelated with the completeness or accuracy of person descriptions. (See **Grass & Sporer, 1991**; **Sporer 2007**; **Wells & Leippe, 1981** - the latter actually found a non-significant negative correlation)

B. **Lineup instructions**: What is the witness told about the composition of the lineup?
 1. **Biased**: E.g., "See if you recognize the offender in the lineup"; do not explicitly indicate that the offender might not be present.
 2. **Unbiased**: Specifically indicate that the offender might not be present in the lineup though, as **Memon et al (2004)** showed, 90% of participants still assume the offender is present.
 3. Effect is primarily on choice probability:
    a. In **target-absent** lineups, biased instructions increase probability of false alarms.
    b. In **target-present** lineups, biased instructions increase the probability of a hit, and the probability of a false alarm (since misses and correct rejections go down.).
 4. Social factors:
    a. Just being asked to identify suspect from lineup suggests presence of the offender in the lineup, and may increase choosing probability. **Wells (1993)** found that if we remove the most commonly identified person from lineup, subsequent choices evenly distributed among the remainder - suggesting increased choice behavior.
    b. **Memon, Gabbert, & Hope (2004)**: Reported that 90% of lineup participants assumed the perpetrator was in the lineup, even though given unbiased instructions that he might not be.

C. **Lineup composition**: Who's in the lineup as foils? How many are there? How are they chosen?
 1. **How many?** Numbers range from 6-12 in different places, and **Nosworthy & Lindsay (1990)** suggest that 3 foils plus suspect is the minimum; increases beyond that do not increase accuracy. (This assumes that these are all plausible foils.)
 2. **How are they chosen**? Two accepted criteria for choosing foils to make sure the lineup is fair (i.e., that a non-witness, given the witnesses description of the offender, would be equally likely to choose any member of an suspect-absent lineup.)
    a. **Match-to-description**: Choose foils so that they match the witness's description of the offender. This is the more common criterion for foil choice. But in a target-absent lineup, there is still the possibility that somebody other than the suspect is

more similar to the offender's undescribed characteristics than the absent offender - this is a problem.
   b. **Match-to-suspect**: Choose foils so that they match the suspect closely. This eliminates the problem of undescribed similarities, but might increase either hits or false alarms. The research to date does not allow us to decide which.

D. **Lineup presentation mode**: Simultaneous vs sequential presentation of lineup members.
   1. **Simultaneous presentation**: The usual one, and always the one depicted on TV. Witness sees the suspect and all the foils at once. Problem is that witness then makes relative judgement - which person looks most like their mental image of the offender. Leads to more false alarms in both suspect-present and suspect absent lineups. (See **Wells, 1984**).
   2. **Sequential presentation**: Witness sees the suspect and the foils one at a time, and has to make a judgement on each one as they appear. In this mode, false alarms in target-absent lineups are reduced (**Lindsay & Wells, 1985**). There seems to be no change in hits from target-present lineups with this procedure - though some studies have suggested a reduction. (See **Steblay et al, 2001**).
      a. Choice tor rejection must occur for each individual in lineup before the next is shown.
      b. Any choice terminates the lineup.
      c. Witness does not know how many people there are in the lineup.
   3. Alternatives to simultaneous or sequential lineups: Might reduce errors.
      a. Separate lineups for different aspects of offender; voice, hair, clothes, etc. (See **Pryke et al, 2004**)
      b. Elimination lineup (especially for children): witness has to successively eliminate individuals before making an absolute judgement about the remaining person (See **Pozzulo & Lindsay, 1999**)

E. **Characteristics of the identification decision**
   1. **Identification latency**: Not tons of research, but generally accurate identifications are faster than erroneous ones. (E.g., **Smith et al, 2001**)
      a. Correlations between accuracy and latency range from -.20 to -.40.
      b. Some indication that specific latency ranges might be a reasonable predictor of accuracy
         (1) **Smith et al (2000)**: 69% accuracy with latencies less than 16 seconds, and 43% for latencies between 16 and 30 seconds, 18% for latencies greater than 30 secs.
         (2) **Dunning & Perretta (2002)**: Suggested a latency window of 10-12 seconds best for accuracy.
         (3) But **Weber et al (2004)** suggest that range for accuracy is broader, between 5 and 30 seconds. But found that for identifications made with high confidence (90%+) made in less than 10 seconds had 88% accuracy
   2. **Identification confidence**:
      a. Want the confidence-accuracy relation (CA) to be **calibrated**: I.e., ideally 90% of identifications made with 90% confidence should be correct, as should 50% of identifications made with 50% confidence, etc.

        b. Research finds such a relationship for choosers, but not for non-choosers (B&W, p. 209)
        c. Appears that in multiple-trial tests, the CA relations most closely approximates the ideal if the proportion of target-absent trials is between 15 and 25%. Target-absent proportions as high as 50% lead to overconfidence.

  F. **Post-identification test variables**: How can events that happen between the identification and the testimony affect the latter?
    1. Post-identification feedback (See **Wells & Bradfield, 1998**; **Wells et al, 2003**). Feedback confirming the subject's identification, whether from police or other witnesses, increases identifier's confidence; disconfirming feedback reduces it. This is important because juries use witness confidence as an indicator of accuracy.
    2. Estimates of witness confidence should be taken and the time of identification; later estimates by the witness should be discounted.
    3. Effects of confirming feedback are less if the witness is already confident, but effect of disconfirming feedback is larger.

XII. Improvements to lineup procedures based on research: **(U.S. Department of Justice Report, 1999)**. About the same recommendations for live and photo lineups.

  A. For Live lineups: (selected recommendations, not all - put all on web?)
    1. Only one suspect per lineup
    2. Match foils to witness description
    3. Minimum of four foils
    4. Give unbiased instructions that offender may not be present
    5. Instruct witness that procedures require that he/she be asked about confidence
    6. Have police present unaware of who the real suspect in the lineup is (not a recommendation, but useful. Not yet common.)

XIII. Within-subject correlations between confidence and memory accuracy are moderate to high, even though the between-subject correlations are low: So while there is no reason to trust a confident source over a less confident source, we can trust the memories in which a witness is more confident vs less confident.

XIV. Brewer et al (2012). Developed and tested new method of lineup identification.

  A. Note several weaknesses with traditional identification test or lineups:
    1. Require witness to make single decision about lineup, either makes a choice from the lineup (i.e., picks the suspect or a foil who is known to be innocent) or rejects lineup

(i.e., says the culprit is not present or cannot decide).
2. Factors like limited exposure to culprit and delay in the identification test mitigate against accuracy of witness memory, so witness has to weigh whether positive or negative decision more appropriate on basis of limited evidence available.
3. Decision criteria may be influenced by social cues (Brewer & Palmer, 2010; Wells et al., 2006) that bias witness toward positive identification.
4. Also, metacognitive cues (e.g., "I saw him for a long time") may persuade witnesses that they have a strong memory and should therefore set strict decision criterion demanding strong evidence for making an identification because the target stimulus should be familiar (Morrell, Gaitan, & Wixted, 2002).
5. Also, perceived importance of identification may bias criterion setting. E.g., may be concerned that identification will result in long jail sentence, so must be correct; or that failing to pick someone might leave dangerous criminal on the streets.

B. Developed technique that required participants to make rapid confidence decisions about a series of 12 faces who might or might not have appeared in a short film they had viewed previously. All foils chosen to match free-recall description of individual given by independent observers. In each of several studies, Ps watched two (or four) short films about which they were going to be asked questions, then watched a distractor film (a TV program) for 5 minutes. In a final experiment, Ps watched four movies (no distractor), and returned to the lab 1 week later.
  1. Two conditions:
     a. Deadline condition: Each face on screen for 3 seconds, with warning buzzer after 2 sec., that judgment had to be made within one second. P rated each face as possible culprit from 100% ("absolutely certain this is the culprit") to 0% ("absolutely certain this is not the culprit").
     b. Control condition: Same as Deadline, but with unlimited time to view each face, and a Yes/No decision required for each face, with lineup ending as soon as a "Yes" decision made. After "Yes", asked to rate confidence in final choice on same 11-point scale as Deadline Ps.
  2. Results:
     a. In first three experiments, with judgements made immediately after viewing films, accuracy significantly higher in every case for Ps in the Deadline condition than in the control condition, using a somewhat complicated accuracy heuristic that combined data across all Ps when a single maximum value did not exist.
     b. Accuracy in the Control condition was .51, .59, and .38 for Expts. 1, 2, amd 3, respectively. (Note the low value for a 1-week delay).
     c. In Deadline condition, individual accuracy much better when difference between highest and next highest confidence rating was .40 or higher.

Table 2. Proportion of Correct Decisions and Number of Decisions for Each Category of Discrepancy Between the Maximum and the Next-Highest Confidence Value in the Deadline Condition

Discrepancy	Experiment 1		Experiment 2		Experiment 3	
	Proportion correct	Number of decisions	Proportion correct	Number of decisions	Proportion correct	Number of decisions
100	1.00	8	.97	33	1.00	3
≥ 90	1.00	11	.89	46	1.00	5
≥ 80	.94	16	.81	64	.83	6
≥ 70	.90	20	.80	83	.70	10
≥ 60	.70	30	.80	96	.64	14
≥ 50	.64	42	.69	128	.68	19
≥ 40	.55	69	.66	147	.56	25
≥ 30	.51	106	.56	202	.47	38
≥ 20	.44	164	.51	270	.42	57
≥ 10	.37	258	.40	388	.34	104

Note: Confidence was rated on a scale from 0% to 100%.

References

Anderson, R.C. & Pichert, J.W. recall of previously unrecallable information following a shift in perspective. *Journal of Verbal Learning and Verbal Behavior*, 1978, 17, 1-12.

Bahrick, H.P., Hall, L.K. & Dunlosky, J. Reconstructive processing of memory content for high versus low test scores and grades. *Applied Cognitive Psychology*, 1993, 7, 1-10.

Bahrick, H.P., Hall, L.K. & Berger, S.A.,. Accuracy and distortion in memory for high school grades. *Psychological Science*, 1996, 7, 265-271.

Bennett, P. Face recall: A police perspective. *Human Learning*, 1986, 197-202.

Bergman, E.T. & Roediger, H.L. Can Bartlett's repeated reproduction experiments be replicated? *Memory and Cognition*, 1999, 27, 937-947.

Bernstein, Thomas, A.K. & Loftus, E.F. Creating bizarre false memories through imagination. *Memory and Cognition*, 2002, 30, 423-431.

Blake, A.B., Nazarian, M. & Castel, A.D. (2015). The Apple of the mind's eye: Everyday attention, metamemory, and reconstructive memory for the Apple logo. *The Quarterly Journal of Experimental Psychology*, published online, February, 2015.

Boon, J.W.C. & Davies, G. Extra-stimulus influences on witness perception and recall: Hastorf and Cantrill revisited. *Legal and Criminological Psychology*, 1996, 1 (Part 2), 155-164.

Boon, J.W.C. & Davies, G. Attitudinal influences on witness memory: Fact and function. In Gruneberg, M.M. (Ed.) *Practical Aspects of Memory: Current Research and Issues.* (Vol., 1, pp. 53-58). Wiley, 1988.

Bornstein, B.H. Liebel, L.M, & Scarberry, M.C. Repeated testing in eyewitness memory: A means to improve recall of a negative emotional event. *Applied Cognitive Psychology*, 1998, 12, 119-131.

Brace, N.A., Pike, G.E. & Kemp, R.I. Investigating E-fit using famous faces. In A. Czerederecka, T. Jaskiewicz-Obydzinska & J. Wojcikiewicz (Eds.) *Forensic Psychology and Law: Traditional questions and new ideas.* Krakow; Institute of Forensic Research Publishers, 2000.

Braun, K.A., Ellis R. & Loftus, E.F. Make my memory: How advertising can change our memories of the past. *Psychology and Marketing*, 2002, 19, 1-23.

Brewer, N., Weber, N., Wootton, D. & Lindsay, D.S. (2012). Identifying the Bad Guy in a Lineup Using Confidence Judgments Under Deadline Pressure. *Psychological Science*, October 2012; vol. 23, 10: pp. 1208-1214.

Brown, A.S. A review of the tip-of-the-tongue experience. *Psychological Bulletin*, 1991, 109, 204-223.

Brown, E., Deffenbacher, K. & Sturgil, W. (1977). Memory for Faces and the Circumstances of Encounter. *Journal of Applied Psychology*, 62(3), 311-318.

Bruce, V., Ness, H., Hancock, P. J. B., Newman, C., & Rarity, J. (2002). Four heads are better than one: Combining face composites yields improvements in face likeness. *Journal of Applied Psychology*, 87, 894–902.

Chandler, C.C. Specific retroactive interference in modified recognition tests: Evidence for an unknown cause of interference. *Journal of Experimental Psychology: Learning, Memory, and Cognition*, 1989, 15, 256-265.

Chandler, CC. How memory for an event is influenced by related events: Interference in modified recognition tests. *Journal of Experimental Psychology: Learning, Memory, and Cognition*, 1991, 17, 115-125.

Christiaansen, R.E. & Ochalek, K. Editing misleading information from memory: Evidence for the coexistence of original and postevent information. *Memory and Cognition*, 1983, 467-475.

Christie, D. & Ellis, H. Photofit constructions versus verbal descriptions of faces,. *Journal of Applied Psychology*, 1981, 66, 358-363.

Clare, J. and Lewandowsky, S. Verbalizing facial memory: Criterion effects in verbal overshadowing. *Journal of Experimental Psychology: Learning, Memory, and Cognition*, 2004, 30, 739-755.

Clark & Tunnicliff (2001). Selecting Lineup Foils in Eyewitness Identification Experiments: Experimental Control and Real-World Simulation. Law and Human Behavior, 2001, 25(3), pp. 199-216.

Clifford, B.R.& Hollin, C.R. Effects of the type of incident and the number of perpetrators on eyewitness memory. *Journal of Applied Psychology*, 1981, 66, 364-370.

Davies, G.M., van der Willik, P. & Morrison, L. Facial composite production: A comparison of mechanical and computer-driven systems. *Journal of Applied Psychology*, 2000, 85, 119-124.

Davies, G.M. & Oldman, H. The impact of character attribution on composite production: A real world effect? *Current Psychology*, 1999, 18, 128-139.

Davies, G.M. & Valentine, T. Facial composites: Forensic utility and psychological research. In Lindsay et al. *The Handbook of Eyewitness Psychology: Volume II, Memory for People*. 2007. Lawrence Erlbaum.

Dunning, D. & Perretta, S. Automaticity and eyewitness accuracy: A 10- to 12-second rule for distinguishing accurate from inaccurate positive identifications. *Journal of Applied Psychology*, 2002,

87, 951-962.

Dunning, D. & Stern, L.B. examining the generality of eyewitness hypermnesia: A close look at time delay and question type. *Applied Cognitive Psychology*, 1992, 6, 643-657.

Ellis, H. Face recall: A psychological perspective. *Human Learning*, 1986, 5, 189-196.

Ellis, H., Shepherd, J. & Davies, G. An investigation of the use of the Photofit technique for recalling faces. *British Journal of Psychology*, 1975, 66, 29-37.

Ellis, H., Davies, G. & Shepherd, H. A critical examination of the Photofit system for recalling faces. *Ergonomics*, 1978, 21, 297-307.

Ellis, H. D., Davies, G. M., & Shepherd, J. W. (1978). Remembering pictures of real and unreal faces: Some practical and theoretical considerations. *British Journal of Psychology*, 69, 467–474.

Eugenio, P., Buckhout, R., Kostes, S. & Ellison, K.W. Hypermnesia in the eyewitness to a crime. *Bulletin of the Psychonomic Society*, 1982, 19, 83-86.

Fiedler, K., Walther, E., Armbruster, T., Fay, D., & Naumann, U. Do you really know what you have seen? Intrusion errors and presuppositions effects on constructive memory. *Journal of Experimental Social Psychology*, 1996, 32, 484-511.

Finger, K. Mazes and music: Using perceptual processing to release verbal overshadowing. Applied Psychology, 2002, 16(8), 887-896.

Flin, R.H. & Shepherd, J.W. Tall stories: Eyewitnesses' ability to estimate height and weight characteristics. *Human Learning*, 1986, 5, 29-38.

Frowd, C. D., Carson, D., Ness, H., McQuiston-Surrett, D., Richardson, J., Baldwin, H., et al. (2005). Contemporary composite techniques: The impact of a forensically-relevant target delay. *Legal and Criminological Psychology*, 10, 61–83.

Frowd, C. D. & Fields, S. (2011): Verbalization effects in facial composite production, *Psychology, Crime & Law*, 17:8, 731-744.

Garry, M. Manning, C.G. Loftus, E.F. & Sherman, S.J. imagination inflation: Imagining a childhood event inflates confidence that it occurred. *Psychonomic Bulletin & Review*, 1996, 3, 208-214.

Garry, M. & Wade, K.A. Actually, a Picture Is Worth Less Than 45 Words: Narratives Produce More False Memories Than Photographs Do. *Psychonomic Bulletin & Review*, 2005, 12 (2), 359-366.

Garry, M., Manning, C.G., Loftus, E.F. & Sherman, S.J. Imagination inflation: Imagining a childhood event inflates confidence that it occurred. *Psychonomic Bulletin and Review*, 1996, 3, 208-214.

Garry, M., & Polaschek, D.L.L. Imagination and memory. *Current Directions in Psychological*

Science, 2000, 9, 6–10.

Goff. L. & Roediger, H. Imagination inflation for action events: Repeated imaginings lead to illusory recollections. *Memory and Cognition*, 1998, 26, 20-33.

Goodsell, C.A., Gronlund, S.D. & Neuschatz, J.S. (2015). Investigating mug shot commitment. *Psychology, Crime & Law*, Vol. 21, No. 3, 219-233.

Grass, E. & Sporer, S.L. Paper presented at German conference, 1991.

Gronlund, S.D. (2004). Sequential Lineups: Shift in Criterion or Decision Strategy. *Journal of Applied Psychology*, 89(2), 362–368.

Gronlund, S.D., Carlson, C.A., Neuschatz, J.S., Goodsell, C.A., Wetmore, S.A., Wooten, A. & Graham, M. (2012). Showups versus lineups: An evaluation using ROC analysis. *Journal of Applied Research in Memory and Cognition*, 1, 221–228.

Gronlund, S.D. & Neuschatz, J.S. (2014). Eyewitness identification discriminability: ROC analysis versus logistic regression. *Journal of Applied Research in Memory and Cognition 3*, 54–57.

Gronlund, S.D., Wixted, J.T., & Mickes, L. (2014). Evaluating Eyewitness Identification Procedures Using Receiver Operating Characteristic Analysis. *Current Directions in Psychological Science*, 23(1) 3–10.

Harris, R. J. Answering questions containing marked and unmarked adjectives and adverbs. *Journal of Experimental Psychology*, 1973, 97, 399-401.

Hastorf, A.H. & Cantrill, H. They saw a game: A case study. *Journal of Abnormal and Social Psychology*, 1954, 49, 129-134.

Heaps, C.M. & Nash, M. Comparing recollective experience in true and false autobiographical memories. *Journal of Experimental Psychology: Learning, Memory and Cognition*, 2001, 27, 920-930.

Heaton-Armstrong, A., Shepherd, E., Gudjonsson, G. & Wolchover, D.(Eds.) *Witness Testimony: Psychological, Investigative and Evidential Perspectives*. Oxford University Press, 2006.

Hopel, L., Lewinski, W., Dixon, J., Blocksidge, D. & Gabbert, F. (2012) Witnesses in Action : The Effect of Physical Exertion on Recall and Recognition. *Psychological Science*, 23(4) 386–390.

Hyman, I.E. & Pentland, J. The role of mental imagery in the creation of false childhood memories. *Journal of Memory and Language*, 1996, 35, 101-117.

Hyman, I.I, Husband, T.H. & Billings, F.J. False memories of childhood experiences. *Applied Cognitive Psychology*, 1995, 9, 181-197.

Keast, A., Brewer, N. & Wells, G.L. Children's metacognitive judgments in an eyewitness identification task. *Journal of Experimental Child Psychology*, 2007, 97(4), Aug., pp. 286-314.

Kitson, A., Darnbrough, M. & Shields, E. Let's face it. *Police Research Bulletin*. 1978, 30, 7-13.

Koriat, A., Goldsmith, M., & Pansky, A. Toward a psychology of memory accuracy. *Annual Review of Psychology*, 2000, 51, 481-537.

Kovera, M. B., Penrod, S. D., Pappas, C., & Thill, D. L. (1997). Identification of computer-generated facial composites. *Journal of Applied Psychology*, 82, 235–246.

Kramer, T.H., Buckhout, R. & Eugenio, P. Weapon focus, arousal, and eyewitness memory. *Law and Human Behavior*, 1990, 14(2), 167-184.

Kuehn, L.L. Looking down a gun barrel: Person perception and violent crime. *Perceptual and Motor Skills*, 1974, 39, 1159-1164.

Light, L.L., Kayra-Stewart, F. & Hollander, S. Recognition memory for typical and unusual faces. *Journal of Experimental Psychology: Human Learning and Memory*, 1979, 5, 212-228.

Lindsay, R.C.L. & Wells, G.L. improving eyewitness identifications from lineups: Simultaneous versus sequential lineup presentations. *Journal of Applied Psychology*, 1985, 70, 556-564.

Lindsay, R.C.L, Martin, R. & Webber, L. Default values in eyewitness descriptions: A problem for the match-to-description lineup foil selection strategy. *Law & Human Behavior*, 1994, 18, 527-541.

Loftus, E. F. *Eyewitness Testimony*. Harvard University Press, 1979.

Loftus, E.F. & Palmer, J. Reconstruction of automobile destruction: An example of the interaction between language and memory. *Journal of Verbal Learning and Verbal Behavior*, 1974, 13, 585-589. 556-564.

Loftus, E.F. & Zanni, G. Eyewitness testimony: The influence of the wording of a question. *Bulletin of the Psychonomic Society*, 1975, 5, 86-88.

Loftus, E.F. Leading questions and the eyewitness report. *Cognitive Psychology*, 1975, 7, 560-572. 556-564.

Loftus, E.F. Made in memory: Distortions in recollection after misleading information. *The Psychology of Learning and Motivation*, 1991, 27, 187-215.

Loftus, E.F., Miller, D. & Burns, H. Semantic integration of verbal information into visual memory. *Journal of Experimental Psychology*, 1978, 4, 19-31.

Loftus, , E.F. , Donders, K., Hoffman, H. & Schooler, J. Creating new memories that are quickly assessed and confidently held. *Memory & Cognition*, 1989, 17, 607-616.

Loftus, E.F., Loftus, G.R. & Messo, J. Some facts about "weapon focus". *Law and Human Behavior*, 1987, 11, 55-62.

Loftus, E.F. & Pickrell, J. The formation of false memories. *Psychiatric Annals*, 1995, 25, 720-725.

Loftus, E.F.. Loftus G.R., & Messo, J. Some facts about "weapon focus". *Law and Human Behavior*, 1987, 11, 55-62.

Loftus, E.F, Miller, D.G. & Burns, H.J. Semantic integration of verbal information into a visual memory. *Journal of Experimental Psychology: Human Learning and Memory*, 1978, 4, 19-31.

MacLeod, M. Retrieval-induced forgetting in eyewitness memory: Forgetting as a consequence of remembering. *Applied Cognitive Psychology*, 2002, 16, 135-149.

MacLin, O.H., Maclin, M. K. & Malpass, R. S. Race, arousal, attention, exposure and delay: An examination of factors moderating face recognition. *Psychology, Public Policy & Law*, 2001, 7(1), 134-152.

Malpass, R.S. & Devine, P.G. Effective size and defendant bias in eyewitness identification lineups. *Law and Human Behavior*, 1981, 5(4), 299-309.

Mazzoni, G. & Memon, A. Imagination can create false autobiographical memories. *Psychological Science*, 2003, 14, 186-188.

McAllister, H.A., Stewart, H.A. & Loveland, J. Effects of mug book size and computerized pruning on the usefulness of dynamic mug book procedures. *Psychology, Crime, & Law*, 2003, 9, 265-278.

McCloskey, M. & Zaragosa, M. Misleading postevent information and memory for events: Arguments and evidence against memory impairment. *Journal of Experimental Psychology: General*, 1985, 114, 1-16.

Meissner, C.A. & Brigham, J.C. Thirty years of investigating the own-race bias in memory for faces: A meta-analytic review. *Psychology, Public Policy, and Law*, 2001, 7, 3-35.

Memon, A., Hope, L., Bartlett, J. & Bull, R. (2002). Eyewitness recognition errors: The effects of mugshot viewing and choosing in young and old adults. *Memory & Cognition*, 30 (8), 1219-1227.

Memon, A., Gabbert, F. & Hope, I. The ageing eyewitness. In J. Adler (Ed.) *Forensic Psychology: Debates, Concepts and Practice*. Devon, UK: Willan. 2004.

Neisser, U. & Harsch, N. Phantom flashbulbs: False recollections of hearing the news about Challenger. In E. Winograd & U. Neisser (Eds), *Affect and Accuracy in Recall: Studies of 'Flashbulb Memories'*. Cambridge University Press, 1992.

Nosworthy, G.J. & Lindsay, R.C.L. Does nominal lineup size matter? *Journal of Applied Psychology*, 1990, 75, 358-361.

Penrod, S. & Cutler, B. (1995). Witness confidence and witness accuracy: Assessing their forensic relation. *Psychology, Public Policy, & Law Special Issue: Witness Memory and Law,* 1, 817, 2013;845

Poole, D.A. & White, L.T. Two years later: Effect of question repetition and retention interval on the eyewitness testimony of children and adults. *Developmental Psychology*, 1993, 29, 844-853.

Porter, S., Yuille, J.C. & Lehman, D.R. The nature of real, implanted, and fabricated memories for emotional childhood events: Implications for the recovered memory debate. *Law and Human Behavior*, 1999, 23, 517-537.

Pozzulo, J.D. & Lindsay, R.C.L. Identification accuracy of children vs adults: A meta-analysis. *Law and Human Behavior*, 1998, 22, 549-570.

Pozzulo, J.D. & Lindsay, R.C.L. Elimination lineups: An improved identification procedure for child eyewitnesses. *Journal of Applied Psychology*, 1999, 84, 167-176.

Pryke, S., Lindsay, R.C.L., Dysart J.E., & Dupuis, P. Multiple independent identification decisions: A method of calibrating eyewitness identifications. *Journal of Applied Psychology*, 2004, 89, 73-84.

Read, J.D. The availability heuristic in person identification: The sometimes misleading consequences of enhanced contextual information. *Applied Cognitive Psychology*, 1995, 9, 91-121.

Roediger, H.L. McDermott, K.B. & Goff, L.M. Recovery of true and false memories: Paradoxical effects of repeated testing. In M.A. Conway (Ed.) *Recovered Memories and False Memories* (pp118-149). Oxford University Press, 1997.

Ross, D.R., Ceci, S.J., Dunning, D. & Toglia, M.P. Unconscious transference and mistaken identity: When a witness identifies a familiar but innocent person. *Journal of Applied Psychology*, 1994, 79, 918-930.

Schacter, D.L. The seven sins of memory. *American Psychologist*, 1999, 54, 182-203.

Schmolck, H., Buffalo, E.A. & Squire, L.R. Memory distortions develop over time: Recollections of the O.J. Simpson trial verdict after 15 and 32 months. *Psychological Science*, 2000, 11, 39-45.

Schooler, J.W. Verbalization produces a transfer inappropriate processing shift. Applied Psychology, 2002, 16(8), 989-997.

Scrivner, E. & Safer, M.A. Eyewitnesses show hypermnesia for details about a violent event. *Journal of Applied Psychology*, 1988, 73, 371-377.

Shapiro, P.N. & Penrod, S. Meta-analysis of facial identification studies. *Psychological Bulletin*, 1986, 100, 139-156.

Shaw, J.S., Bjork, R.A. & Handal, A. Retrieval-induced forgetting in eyewitness-memory paradigm. *Psychonomic Bulletin and Review*, 1995, 2, 249-253. Available]

Simmons, D. & Chabris, C. Gorillas in our Midst: Sustained inattentional blindness for dynamic events. *Perception*, 1999, 28, 1055-1074.

Simons, D.J., & Levin, D.T. Failure to detect changes to people in a real-world interaction. *Psychonomic Bulletin and Review*, 1998, 5(4), 644-649.

Sirovich, L. & Kirby, M. Low dimensional procedure for the characterization of human faces. *Journal of the Optical Society of America A*, 1987, 4, 519-524.

Smith, S.M. & Vela, E. Environmental context-dependent eyewitness recognition. *Applied Cognitive Psychology*, 1990, 6, 125-139.

Smith, S.M., Lindsay, R.C.L. & Pryke, S. Predictors of eyewitness errors: Can false identifications be diagnosed? *Journal of Applied Psychology*, 2000, 85, 542-550.

Smith, S.M., Lindsay, R.C.L., Pryke, S. & Dysart, N. Postdictors of eyewitness errors: Can false identifications be diagnosed in the cross-race situation? *Psychology, Public Policy and Law*, 2001, 7, 153-169.

Sporer, S.L. *An archival analysis of person descriptions.* Paper presented at the Biennial Meeting of the American Psychology-Law Society, San Diego, CA, 1992.

Sporer, S.L. Person descriptions as retrieval cues: Do they really help? *Psychology, Crime and Law*, 2007, 13(6), 591-609.

Steblay, N.M. A meta-analytic review of the weapon focus effect. *Law and Human Behavior*, 1992, 16, 413-424.

Steblay, N., Dysart, N., Fulero, J. & Lindsay, R.C.L. Eyewitness accuracy rates in sequential and simultaneous lineup presentations: A meta-analytic comparison. *Law and Human Behavior*, 2001, 25, 459-473.

Stewart, H.A. & McAllister, H.A. Simultaneous vs sequential mugbook procedures: Some surprising results. *Journal of Applied Psychology*, 2001, 86, 1300-1305.

Talarico, J.M. & Rubin, D.C. Confidence, not consistency, characterizes flashbulb memories. *Psychological Science*, 2003, 13, 455-461.

Technical Working Group for Eyewitness Evidence (1999). *Eyewitness Evidence: A Guide for Law Enforcement.* Washington, DC: U.S. Department of Justice, Office of Justice Programs.

Thomas, A.K., Bulevich, J.B. & Loftus, E.F. Exploring the role of repetition and sensory elaboration in the imagination inflation effect. *Memory and Cognition*, 2003, 31, 630-640.

Tuckey, M.R. & Brewer, N. The influence of schemas, stimulus ambiguity, and interview schedule on eyewitness memory over time. *Journal of Experimental Psychology: Applied*, 2003, 9, 101-118.

Tunnicliff & Clark (2000). Selecting foils for identification lineups: Matching suspects or descriptions? Law and Human Behavior, 2000, 24(2), pp. 231-258.

Turk, M. & Pentland, A. Eigenfaces for recognition. *Journal of Cognitive Neuroscience*, 1991, 3, 71-86.

Valentine, T., Pickering, A., & Darling, S. Characteristics of eyewitness identification that predict the outcome of real lineups. *Applied Cognitive Psychology*, 2003, 17, 969-993.

van Koppen, P. & Lochun, S. Portraying perpetrators: The validity of offender descriptions by witnesses. *Law & Human Behavior*, 1997, 21, 661-685.

Weber, N., Brewer, N., Welles, G.L., Semmler, C., & Keast, A. Eyewitness identification accuracy and response latency: The unruly 10-12 second rule. *Journal of Experimental Psychology: Applied*, 2004, 10(3), 139-147.

Weinberg, Howard I.; Wadsworth, John; Baron, Robert S. Demand and the impact of leading questions on eyewitness testimony. *Memory & Cognition*, 1983, 11(1), 101-104.

Wells, G.L. The psychology of lineup identification. *Journal of Applied Social Psychology*, 1984, 14, 89-103.

Wells, G.L. what do we know about eyewitness identification? *American Psychologist*, 1993, 48, 553-571.

Wells, G.L. & Bradfield, A.L. "Good, you identified the suspect": Feedback to eyewitnesses distorts their reports of the witnessing experience. *Journal of Applied Psychology*, 1998, 83, 360-376.

Wells, G.L. & Bradfield, A.L. Distortions in eyewitnesses' recollections: Can the postidentification feedback effect be moderated? *Psychological Science*, 1999, 10, 138-144.

Wells, G.L. Olson, E.A. Charman, S.D. Distorted retrospective eyewitness reports as functions of feedback and delay. *Journal of Experimental Psychology: Applied*, 2003, 9, 42-52.

Wells, G. L., Charman, S. D., & Olson, E. A. (2005). Building face composites can harm lineup identification performance. *Journal of experimental psychology: Applied*, 11, 147–156.

Wells, G.L. & Leippe, M.R. How do triers of fact infer the accuracy of eyewitness identifications? Using memory for peripheral details can be misleading. Journal of Applied Psychology, 1981, 66, 682-687.

Wells, G. L; Memon, A., & Penrod, S.. Eyewitness Evidence: Improving its Probative Value. D. *Psychological Science in the Public Interest*. Vol 7(2), Nov 2006, pp. 45-75

Wells, G. L., & Olson, E. (2003). Eyewitness identification. *Annual Review of Psychology*, 54, 277–295.

Wiseman, R., Watt, C., ten Brinke, L., Porter, S., Couper, S-L. & Rankin, C. (2012). The Eyes Don't Have It: Lie Detection and Neuro-Linguistic Programming. PloS ONE 7(7). E40259.

Wright, D.B. & Sladden, B. An own gender bias and the importance of hair in face recognition. *Acta Psychologica*, 2003, 114, 101-114.

Wright, D.B. & Stroud, J.N. age differences in lineup identification accuracy: People are better with their own age. *Law and Human Behavior*, 2002, 26, 641-654.

Yuille, J.C. & Cutshall, J.L a case study of eyewitness memory of a crime. *Journal of Applied Psychology, 1986*, 71, 291-301.

Criminal Profiling

I. History of criminal profiling

   A. Criminal profiling began in fiction more than 125 years before it began in reality:
      1. Edgar Allan Poe's detective C. Auguste Dupin in the "Murders in the Rue Morgue" (1841)
      2. Arthur Conan Doyle's Sherlock Holmes (1892-1927), is the model of a forensic scientist who relies on a carefully honed intuition, though not often psychology.

   B. First modern case of serial murder occurs at about this time: the Whitechapel Murders committed by 'Jack the Ripper'.
      1. Murder of five prostitutes between August and November of 1888 prompted first non-fictional profile we by Dr. Thomas Bond, specialist in forensic medicine who performed autopsies on two of the victims. He submitted a report to the head of the London Criminal Investigation Division (CID) in which he speculated about the murderer::
      2. *"The murderer must have been a man of great physical strength and of great coolness and daring. There is no evidence that he had an accomplice. He must in my opinion be a man subject to periodical attacks of Homicidal and Erotic mania. The character of the mutilations indicate that the man may be in a condition sexually, that may be called Satyriasis. .... The murderer in external appearance is quite likely to be a quiet inoffensive looking man probably middle-aged and neatly and respectably dressed. I think he must be in the habit of wearing a cloak or overcoat or he could hardly have escaped notice in the streets if the blood on his hands or clothes were visible. Assuming the murderer to be such a person as I have just described, he would be solitary and eccentric in his habits, also he is most likely to be a man without regular occupation, but with some small income or pension. He is possibly living among respectable persons who have some knowledge of his character and habits and who may have grounds for suspicion that he isn't quite right in his mind at times. Such persons would probably be unwilling to communicate suspicions to the Police for fear of trouble or notoriety, whereas if there were prospect of reward it might overcome their scruples."* (Rumbelow, 1975, p. 138)

II. Purposes and Challenges of Offender Profiling

   A. Why do we do offender profiling? What use is it?
      1. Provide characteristics of the offender
      2. Analyze the crime scene
      3. Provide leads for investigators to follow
      4. Reduce the viable subject pool in a criminal investigation

5. Prioritize the investigation into those suspects
6. Assess risk of offender escalation
7. Provide a psychological evaluation of items found in the possession of suspects
8. Assist in developing interview and/or interrogation strategies
9. Investigate/demonstrate links between crimes
10. Provide supportive trial testimony

B. Assumptions of profiling (Pascal **Chifflet, 2014**)
  1. **Behavioral consistency assumption**: Offenders behave in a generally consistent manner each time they offend.
     a. For this assumption, behaviors must not only be consistent across an offender's crimes, but must, overall, be of relatively low frequency in all such crimes.
     b. Behaviors that are useful in linking crimes differ from one crime type to another.
        (1) **Bateman, A. L., & Salfati, G. (2007)** examined 35 serial homicide behaviors used by 90 offenders in 450 serial homicide cases.
           (a) Several aspects of homicides e.g., bringing a crime kit, destroying evidence) both consistent and low frequency overall, making them ideal for crime linkage analysis.
           (b) Restraining victim and bringing weapon very consistent, but also high in frequency.
        (2) **Bennell & Canter (2002)** examined commercial robbery cases and found inter-crime consistency in several features, especially distances between crimes.
  2. **Behvioral differentiation assumption:** The manner in which a particular offender behaves is distinguishable from that of another offender
  3. **Homology assumption**: Offenders exhibiting similar criminal behaviour will possess similar characteristics
     a. **Doan, B., & Snook, B. (2008)** used existing typologies to classify 87 arson cases and 177 robbery cases.
        (1) Then compared background characteristics of criminals who committed various crime types.
        (2) Using pre-existing typologies to classify crimes into mutually exclusive types was difficult.
        (3) Homology assumption violated in 56% of the comparisons of background characteristics between different arson types and 67% of the comparisons between the different robbery types.
     b. **Mokros, A., & Alison, L. J. (2002).** Compared similarity of crime scene actions and similarity of sociol-demographic chcaracteristics in 100 British male stranger rapists.
        (1) Correlationally tested whether increased similarity in offence behaviour coincided with increased similarity in socio-demographic features and previous convictions.
        (2) No significant correlations: Rapists with similar crime scene actions not more similar with respect to age, socio-demographic features, like employment situation, ethnicity, or criminal records.
        (3) Findings provide no support for the assumption of homology between crime

scene actions and background characteristics for rapists in the sample.
- c. **Doan, B., & Snook, B. (2008).** Also failed to find evidence for the homology assumption in a sample of robbery and arson cases.
  - (1) Using existing typologies, classified 87 arsons and 177 robberies into different crime types, then compared background characteristics of offenders.
  - (2) Found that pre-existing typologies to classify crimes into mutually exclusive types not easily accomplished.
  - (3) Notwithstanding homology assumption violated in 56% of the comparisons of background characteristics between the different arson types and in 67% of the comparisons of background characteristics between the different robbery types.
  - (4) Overall, 73% of the effect sizes for the associations between crime type and background characteristics were low to moderate ($V < .3; d < .2$).
- d. **Alison, L. J., Bennell, C., Mokros, A., & Ormerod, D. (2002).** Argue that the homology assumptions has received little support because it relies on a purely trait-based model of personality, that does not take into account interations between the person and the situation.
- e. **Canter et al (2004)** could find no basis for the organized/disorganized distinction in serial murder, though it forms the basis of the FBI's widely-used crime/criminal typology.
  - (1) FBI dichotomy based on limited sample of 36 serial murderers who agreed to talk to the FBI profilers. The interviews were unstructured, and questions differed from one interviewee to the next, and there was no attempt to validate the distinction on another sample.
  - (2) Examined 100 U.S. serial murder cases and found no evidence that organized features were more highly correlated with each other than with disorganized features or actions.
- f. **Canter & Wentick (2004):** empirically tested the Holmes and Holmes's serial murder classification scheme using crime scene evidence from 100 U.S. serial murders.
  - (1) The co-occurrence of content categories derived from the crime scene material was submitted to smallest space analysis.
  - (2) Features characteristic of "power or control" killings found to be typical of the sample as a whole, occurring in more than 50% of cases, and thus did not form a distinct type.
  - (3) Limited support found for aspects of the lust, thrill, and mission styles of killing, but only in differences in the ways victims were dealt with (e.g., mutilation, restraints, ransacking their property), not with respect to the motivations implicitly inferred in the Holmes and Holmes's typology.

C. What are the challenges of offender profiling?
  1. Think about how difficult this is: Take info from crime scene - not the individual - and infer from it a psychological profile that gives us the person's age, personality, work habits, interests, character flaws, etc.
  2. Even modern personality test, given to the person, don't allow us to do this.
  3. Mostly a matter of intuition and experience, not science at all

4. Not thoroughly evaluation - not at all clear how useful the exercise is.

III. History of Modern profiling:

A. Psychiatrist Dr James A. Brussel made first systematic offender profile of person responsible for a series of New York bombings from 1940 to 1956.
   1. Profile: "Male, former employee of Consolidated Edison, injured while working there so seeking revenge, paranoid, 50 years old, neat and meticulous persona, foreign background, some formal education, unmarried, living with female relatives but not mother who probably died when he was young, upon capture he will be wearing a buttoned up double breasted jacket. "
   2. Reasoning behind profile:
      a. Common sense: E.g. male (like vast majority of bombers).
      b. Former employer Consolidated Edison: Based on content of bombers letters.
      c. Intuitive (Sherlock Holmesian): Foreign birth based on stilted writing (e.g. 'dastardly deeds') and absence of contemporary slang.
      d. Brussel believed bomber had 'Oedipal complex', so unmarried and living with female relatives. Also noted 'phallic' construction of bombs, 'breast-like' W's in hand written letters, and tendency to 'slash' and 'penetrate' the seats when planting bombs in movie theaters.
   3. ConEd staff search employee files for anyone matching profile, and find George Metesky - filed unsuccessful disability claim after accident at work. Wrote letters to company, one of which mentioned 'dastardly deeds'. When arrested, Metesky confessed and was taken to police station wearing predicted buttoned up double breasted jacket.

B. FBI's Behavioral Analysis Unit
   1. Formed in 972 as Behavioral Science Unit (BSU)
   2. Now called Behavioral Analysis Unit, and is part of **National Center for the Analysis of Violent Crime** (NCAVC):
      a. **Behavioral Analysis Unit** (BAU)
      b. **Child Abduction Serial Murder Investigative Resources Center** (CASMIRC)
      c. **Violent Criminal Apprehension Program** (VICAP): Nationwide data information center for collection, collating, and analysis of violent crimes - specifically murder. Cases examined include homicides, especially those involving abduction, that seem random, motiveless, or sexually oriented; or are known or suspected to be part of a series. Facilitates cooperation, communication, and coordination between law enforcement agencies and provides support in investigation and apprehension of violent serial offenders.

C. Profiling received higher profile with 1993 release of "Silence of the Lambs" based on book by Thomas Harris, who had visited the BAU as part of his prep for the book, as based his depiction of Agent Jack Crawford on John E. Douglas.

IV. FBI's Model of Profiling

   A. Introduction
      1. First published description of a model of profiling.
      2. 6 - Stage model whose terminology still employed in law enforcement,

   B. **Stage 1, Profiling inputs**: Gather evidence and organize it into categories:
      1. **Crime scene information**:
         a. **Physical evidence**: E.g., blood spatters, footprints, tools and other objects left at the scene, info re ease of access to the crime scene, weather and traffic patterns in the area, social and political climate in the area
         b. **Pattern of evidence**: Integration of physical evidence into a synopsis of the crime
         c. **Weapons**: Murder weapon and any other weapons used in the commission of the crime, e.g., those used to control of subdue victim
      2. **Victimology**: Who was the victim?
         a. Age, physical condition, occupation; personality, criminal record; habits, hobbies, social conduct, relationships with family; where seen last
      3. **Forensic information**: Time and cause of death; sequence of wounds; evidence of sexual assault; toxicology report
      4. **Preliminary police reports**: Observations about the scene, presence of witnesses; physical and socioeconomic description of crime neighborhood; crime scene photos.

   C. **Stage 2, Decision-Process Models**: Integrate info from profiling inputs into crime classification:
      1. **Homicide type and style**:
         a. **Single, double, or triple homicide**: One, two, or three victims killed as part of a single event.
         b. **Mass murder**: Four or more victims in one location. 'Classic' mass murder involves one killer, one location, murders over minutes, hours, or days. "...usually a mentally disordered individual who releases his frustration and hostility by acting violently against a group of people otherwise unrelated to him." (Hicks & Sales, p. 20)
         c. **Spree murder**: Multiple killings at more than one location with no emotional 'cooling off' period between them. Killings all part of a single event.
         d. **Serial murder**: Three of more victims in three or more separate events with an emotional cooling-off period between them. "Typically premeditated and planned" (H&S, p. 20)
      2. **Victim Risk**: Those factors that might have led the victim to be targeted, e.g., age, occupation lifestyle, size, strength, location.
      3. **Offender Risk**: How much risk did the offender take in committing the crime?
      4. **Escalation Risk**: info from sequence of behaviors, and from other sections of the Decision-Process Models stage, are used to make this determination.

5. **Time Factors**: How much time did the offender spend with the victim? Did the crime taken place during daylight or after dark? (Latter may be a clue to offender's occupation and/or habits/
6. **Location Factors**: Where was victim approached; where was crime committed. Are they both in the same place? What about transportation, if crime occurred in different place from initial contact with victim.

D. **Stage 3, Crime Assessment**:
  1. **Reconstruction of the crime**: What happened and in what order. How was the crime planned or organized, if it was? How was the victim coerced and/or controlled?
  2. **Crime Classification**: Was the crime and offender *organized* or *disorganized*?
     a. **Victim selection**:
        (1) Organized offenders target particular victim or type of victim
        (2) Disorganized offenders take victims of opportunity, or by chance.
     b. **Control Strategies**:
        (1) Organized: Use deception (con or ruse), but may use force to maintain control.
        (2) Disorganized: Use physical force to coerce and control. Crime scene will often indicate lack of complete control over victim.
     c. **Sequence of crime**:
        (1) Organized: Plan the crime sequence according to their fantasies.
        (2) Disorganized: No fantasy scripts, and sequence is usually unplanned.
  3. **Staging**: Some offenders alter crime scene to mislead investigators, e.g. removing items to make crime appear as a robbery.
  4. **Motivation**: May be easier to determine for organized offenders; often difficult for disorganized offenders.

E. **Stage 4, Criminal Profile**: Information integrated into a profile of the unknown offender - or UNknown SUBject (UNSUB). "Statements in the profile are considered to be hypotheses, and it is not expected that every statement will be accurate." (H&S, p. 23) Predictions will or should be made about:
  1. **Demographics**: Offender's intelligence, education, military experience, occupation and job status, "living circumstances, interpersonal style, relationship to others, and (if applicable) the make and color of his vehicle." (H&S, p. 23)
  2. **Physical characteristics**: Age, height, weight, build, ethnic background.
  3. **Habits**: E.g., "neatness or disorganization, drug and alcohol use; and hobbies and interests such as metalworking, athletics, and so on." (H&S, p. 24)
  4. **Beliefs and Values**: E.g., about women, sexual relations, religion, politics, etc.
  5. **Pre- and Post-offense behavior**: Predictions about, e.g., "any unusual behavior the perpetrator might have exhibited before the commission of the crime, ...about stressors that may have triggered the act, and information about what to look for in the offender's postoffense behavior (e.g., excessive drinking, bragging)." (H&S, p. 24)
  6. **Recommendations**: How to proceed with the investigation.
     a. "One piece of advice might be to plant information or 'traps' in the news media to encourage the perpetrator to reveal himself... An example would be to publicize the funeral or anniversary of the victim's death and then monitor the cemetery for any

unusual visitors." (H&S, p. 24)
- b. Tips for interviewing suspects (e.g., hard or soft style, play to particular emotions or aspects of self, etc.)

F. **Stage 5, Investigation**:

G. **Stage 6, Apprehension**:

V. Profiling Model of **Holmes & Holmes (1996)**

A. Introduction: Like that of Douglas et al, this is represents an intuitive analysis of crime scene evidence. Homes & Holmes describe profiling as *"in part a gift reserved to certain individuals who can reach inside the criminal mind and understand it. "* (p. 166 of Holmes & Holmes)

B. Base their model on matching crime scene evidence to a set of offender typologies:
   1. **Disorganized Asocial vs Organized Nonsocial offenders**:
      a. Distinction comes from **Ressler, Burgess & Douglas (1988)**: Organized offender described as having an unspecified character disorder, and a 'masculine personality'. Is a 'flashy' dresser, and drives a car that 'reflects his personality'.
      b. Disorganized offender is disorganized in all facets of life, and 'asocial' in the sense of being a 'loner'. Does not associate with others since others are not good enough for his company.
   2. **Serial murderer types**: Apparently based on interviews and case studies (Holmes & DeBurger, 1985), but not clear how typology was derived from the interview data.
      a. **Spatial mobility**:
         (1) Geographically stable commit crimes in same, nearby, area
         (2) Geographically transient commit crimes away from home.
      b. **Visionary Serial Killer**: Murders because of visions or voices. Generally declared incompetent or insane in court.
      c. **Mission Serial Killer**: Feel conscious need to eliminate certain group of people (e.g. prostitutes, etc.). Not psychotic; is typical of organized nonsocial offender.
      d. **Hedonistic Serial Killer**: Derives sexual pleasure or personal gain from the crime. If sexual, will often prolong killing o f victim, adding acts of mutilation, torture, dismemberment, etc. Those for personal gain (comfort-oriented serial murder) include paid assassins and those who kill relatives for money.
      e. **Power/Control Serial Killer**: Gets sexual pleasure from exerting power, control, domination over victim. Is said to prolong killing scene and kill 'hands-on', e.g. by strangulation.
      f. All categories describe motive, but are less than useful in narrowing down list of suspects.
   3. **Rapists**: Their model heavily psychodynamic, positing a rejecting, controlling, seductive mother: "the professional literature suggests that parental rejection,

domination, cruelty, and seductiveness are important factors in the early life of the rapist." (Holmes & Holmes, 1996, p. 118) Typology based on research reported by **Groth, Burgess & Holmstrom (1977)**, and elaborated by **Knight & Prentky (1987)**:

    a. **Power reassurance**: "Passive social loner with feelings of inadequacy. Single, nonathletic, lives with parents while working at menial job. Offends to elevate his own self-status." Chooses victims from his neighborhood in early morning hours; travels on foot. Believes victim enjoys the rape, and may contact victim later to ask about their well-being."

    b. **Anger retaliation**: "Wants to hurt women. Socially competent and athletic, with action-oriented occupation. Probably married but hates women and frequents bars. Attacks often unplanned, occur near home, and involve intent to harm."

    c. **Exploitive**: "Believes he is entitled to rape as expression of dominance. Athletic, with macho occupation, drives 'flashy' car and has series of unsuccessful marriages. Frequents bars and may have history of property damage as well as dishonorable discharge from military. Tends to rape in 20- to 25-day cycles. Picks up victims in bars and does not conceal identity from victim. Plans attack, and controls victim with aggression. Sociopathic/psychopathic.

    d. **Sadistic**: Considered most dangerous type. Associates aggression with sexual gratification. Typically married, middle class, in 30s. White collar job and no prior arrests. Stalks victim, uses excessive restraints. Violence of crimes often escalates to murder.

4. Also have typologies for arson and child molestation, which we will not consider.

C. Other Factors Considered or Suggested in the Model:
  1. *Crime location type*: Consider five possible locations related to a murder or rape:
    a. **Encounter site**: Where did offender and victim first meet?
    b. **Attack site**: Where did offender first attack victims? (May be the same as encounter site)
    c. **Crime site**: Where did the murder or rape crime take place? (May be the same as attack and/or encounter site)
    d. **Victim disposal site**: Where victim is released or body dropped.
    e. **Vehicle dump site**: (when relevant)
  2. **Victim Profiling**: H&H think this is important, and is not done well at the moment. Physical Traits (age, sex, hair and eye color, hairstyle, mode of dress); Marital Status; Personal Lifestyle (friends, hobbies, drug use, haunts); Occupation; Education; Personal Demographics (racio-ethnic identity, neighborhood, previous residences); Medical History; Psychosexual History; Criminal Justice System History; Last Activities.

VI. Profiling Model of **Keppel and Walter (1999)**

  A. Designed to overcome or compensate for shortcomings of Holmes & Holmes model:

1. Typologies "have a wide range of function, are of limited service to investigative work, and are unsupported by empirical study). Keppel & Walter, 1999, p. 418.)
2. Rape categories derived from **Groth et al (1977)**; adapted by Douglas et al for Crime Classification Manual, and adopted by Holmes & Holmes (1996), and Turco (1990).
3. Constructed rape-murder typology to compensate for failures in H&H model. Typology contains four categories:

B. **Power-Assertive Rape-Murder**
   1. **Dynamics**: Rapes are planned, but murder is consequence of increased aggression to control victim. Offender tries to demonstrate dominance and mastery over victim by maintaining assertive image and using violence. Killing victim reinforces offender's power by eliminating any threat from the victim. This type "analyzes ways to improve on this macho image and power." (H&S, p. 37)
   2. **Homicidal Pattern**: Hallmark is assertion of power through rape and murder. Often assault of opportunity, with offender bringing own weapon, which he views as an extension of his power.
   3. **Suspect Profile**: Emotionally primitive male in early 20s. Macho bodybuilder and displays tattoos as an expression of masculinity.
      a. Car well kept
      b. May use drugs and alcohol heavily.
      c. Arrogant and condescending attitude, not viewed as a team player.
      d. Interest in athletics restricted to individual contact sports (e.g., wrestling, boxing).
      e. History of burglary, theft, or robbery, but no contact with mental health system.
      f. Likely a school dropout with military service with poor service or early termination
      g. May have had multiple unsuccessful relationships, demonstrates unconventional sexual interests; may be strongly anti-gay.

C. **Power-Reassurance Rape-Murder**:
   1. **Dynamics**: Expresses sexual competence through seduction. Rape is planned, but murder characterized as "planned overkill of the victim." Rapist motivated by seduction-conquest fantasy and may panic when victim refuses to comply with fantasy, leading to assault and murder. May commit post-mortem mutilation out of curiosity.
   2. **Homicidal Pattern**: Victim usually 10-15 years older or younger than offender, and known to him. May initially use threats or a weapon to gain control of victim before acting out the seduction fantasy. Offender may attempt polite verbal dialogue designed to elicit reassurance of his sexual competence. When rejected, feels threatened and kills victim via beating or strangulation. Sexual assault may be incomplete, so no semen at the scene. May attempt to continue relationship with victim by taking souvenir, or collecting newspaper stories about the crime. Assaults are likely to be episodic, and take place at night.
   3. **Suspect Profile**: Usually in mid-20s (though perhaps older if incarcerated during mid-20s). Daydreams and fantasizes obsessively, and therefore appears emotionally scattered. Prefers to live in fantasy rather than risk rejection in sexual relationships, and is therefore likely unmarried. Seen by others as cold and socially isolated. Will have unremarkable educational and military history. May be seen as underachiever, and may have received mental health referral as a result. Feels inferior, unable to

handle criticism. Likely to live at home and work at menial job. Most likely to walk rather than drive, but if has car will be poorly kept older model. Criminal history may include peeping, unlawful entry, larceny. Leaves a disorganized crime scene with lots of evidence.

D. **Anger-Retaliatory Rape-Murder**:
1. **Dynamics**: Rape is planned, and killing characterized by venting of anger, revenge toward victim. Attack may be precipitated by criticism of offender by victim or another woman with power over him. Assaults likely to be episodic and repeated to relieve offender's stress.
2. **Homicidal Pattern**: Sexual assault is violent, with 'overkill'. Victim is typically a substitute for a woman who has criticized or humiliated him. When target is younger than offender, she is likely to be assaulted directly rather than through a substitute. "The killer typically walks to the crime scene. If he drives, he will park and walk the last 200 feet to the crime scene on foot. The victim will be hit in the mouth and face, and the offender may use weapons of opportunity. The rape may not be completed, but the assault will continue until the perpetrator feels emotionally satisfied, regardless of whether the victim is till alive. Postmortem, the body is placed on its side, away from the door, face down, with the eyes covered, or in the closet with the door closed. The crime scene is typically disorganized, with the weapon left within 15 feet of the body. The offender is likely to take a souvenir before leaving the scene. Because the perpetrator blames the victim, he does not experience any feelings of guilt or responsibility. Instead, he may feel sentimental toward the victim and assist in the search for her body" (H&S, p. 39)
3. **Suspect Profile**: Usually in mid- to late-20s, and targets older victims. Seen by others as impulsive, self-centered, temperamental. A longer, with superficial social relationships. If married, estrangement, domestic violence, and/or extramarital affair possible. Sexually frustrated, may be impotent. Criminal history may include assaults, domestic violence, reckless driving. If was in the military, discharge likely. May have previous referrals to mental health professionals.

E. **Anger-Excitation Rape-Murder**:
1. **Dynamics**: Both rape and murder are premeditated. Victim either male or female, and offender gains gratification through inflicting pain and terror via prolonged torture. Assault is driven by offender's fantasies of dominance and control, as well as his interest in the process of killing. "The offender's anger is eroticized and rehearsed through fantasy, and the ultimate intent is one of 'indulgent luxury'" (H&S. P. 40, from Keppel & Walter, p. 431)
2. **Homicidal Pattern**: Pattern reflects a planned and prolonged assault. Offender brings weapons and tools to the crime scene. Victim may be stranger, but generally a preferred type of person. Ruse or con used to lure victim to isolated location. Offender may display changing moods that confuse victim. May also inform victim of intent to kill. Assault "contains elements of ritual and experimentation, characterized by bondage and domination. There may be evidence of cutting, bruises, incomplete strangulation, washing, shaving, and burning. Sexual experimentation continues post-mortem." (H&S, p. 40) Offender is organized and commits crimes away from home.

May attempt to involve himself in the criminal investigation.
   3. **Suspect Profile**: Age varies; appears normal and has lifestyle separate from criminal activities. Likely to be married and works best under minimal supervision. Employment may involve mechanical woks or carpentry. Education and military history will reflect his organization, and may have college education. May also have private room for murder kit, souvenirs, and collection of porn with bondage/sadism themes.

F. Distribution of types in the offender population: K&W looked over records of 2,476 offenders in jail for sexually-related murders, and concluded that:
   1. 38% power-assertive
   2. 34% anger retaliatory
   3. 21% power reassurance
   4. 7% anger excitation
   5. So all offenders fell into their classification system.

VII. Profiling Model of **Turvey (1999)**: Large text proposing **deductive** model of profiling that relies on physical and behavioral evidence to draw inferences about offenders,

A. **Inductive** vs **deductive** profiling methods. We often think of induction as going from the specific to the general. In this case, we are focusing on the statistical aspects of induction: Taking a set of specific observations (statistical) and deriving a general conclusion, and the aspect of induction that the premises may be false because they are generalizations.
   1. Describes **inductive profiling** as "a comparative, correlational, and/or statistical process reliant upon subjective experience" (p. 14) It "involves broad generalizations or statistical reasoning, where it is possible for the premises to be true while the subsequent conclusion is false" (p. 16).
      a. E.g.:
         (1) Premise: The rape victim was a white female.
         (2) Premise: Most rapists commit sexual assaults against persons in their own ethic group.
         (3) Premise: Most rapists have not served in the military.
         (4) Conclusion: Offender is while male with no military service.
      b. Conclusion problematic because the statistics are questionable, and because of inherent weakness in generalizing from nomothetic data to the individual offender.
   2. Describes **deductive profiling** as : "forensic-evidence-based, process-oriented, method of investigative reasoning about the behavior pattern of a particular offender."(p. 14) It "involves conclusions that flow logically from the premises stated. It is such that if the premises are true, then the subsequent conclusion must also be true. (P. 16)
      a. E.g.
         (1) Premise: The offender disposed of his victim's body in a remote area of the mountains.
         (2) Premise: Tire tracks were found at the disposal site.

(3) Conclusion: If the tire tracks belong to the offender, then the offender has access to a vehicle, and is able to be mobile." (P. 27)
　　　b. Argues that while deductive profiling not wholly scientific, it is based on scientific thinking.

B. Four components of the deductive profiling method:
   1. **Forensic and behavioral evidence**: Reconstructing the sequence of events in the crime, including offender's and victim's behavior.
      a. Witness and victim statements
      b. Crime scene photos
      c. Wound pattern analysis
      d. Blood spatter analysis
      e. Ballistics analysis
      f. Other forensic analyses
   2. **Victimology**:
      a. Physical characteristics
      b. Habits and lifestyle
      c. Relationships
      d. Risk level
   3. **Crime scene characteristics**:
      a. Method of attack
      b. Nature and sequence of sexual or violent acts
      c. Verbal behavior
      d. Precautionary acts
   4. **Deduction of offender characteristics** from above: "considerably artful, and therefore a matter of expertise and not science" (p. 31)

C. **Behavior-motivational typology** (derived from **Groth et al, 1977**; adapted by Douglas et al for Crime Classification Manual, and adopted by Holmes & Holmes, 1996 and Turco, 1990). Originally intended to cover rapes only, Turvey applies them more broadly, arguing that "The needs, or motives, that impel human criminal behavior remain essentially the same for all offenders, despite their behavioral expression that may involve kidnapping, child molestation, terrorism, sexual assault, homicide, and/or arson." (P. 170). Applies classification to crime *behavior*, rather than to *offenders*.
   1. **Power reassurance (Compensatory)**: Behaviors designed to restore offender's self-confidence through low levels of aggression. May also manifest themselves in belief that offense is consensual, or that victim is willing. (Turvey, p. 312)
   2. **Power assertive (Entitlement)**: Behaviors designed to restore offender self-confidence using moderate to high levels of aggression. Sense of personal inadequacy expressed through control, mastery, and humiliation of victim. (Turvey, p. 313)
   3. **Anger retaliatory (Anger or Displaced)**: Behaviors suggesting considerable rage toward victim or group, or symbol of either. Usually stranger assaults, domestic assaults, work-related homicides. (Turvey, p. 316)
   4. **Anger excitation (Sadistic)**: Behaviors indicating gratification from victim's pain and suffering. Primary motivation is sexual, but manifested as aggression or even torture. (Turvey, p. 318)

5. **Profit (Material Gain)**: Can be found in all types of homicides, and in thefts, arson, kidnapping, fraud, extortion, etc. (Turvey, p. 320)

D. Turvey is broadly critical of the empirical literature (e.g., polygraphy, geographic profiling, smallest space analysis) as 'scientification', and not real science. Thinks much of profiling is art not science.

VIII. Hicks and Sales (2006) Critique of Nonscientific Profiling Models (H&S, pp. 49-70.)

A. **Lack of goals and standards**:
  1. No agreement on what the goals of profiling are, or should be. Many models state no explicit goals. When they do, there is no clear consensus about what these goals should be.
     a. Some see benefits only in the investigative phase
     b. Others see uses in interrogation, and even in trial strategy
     c. No agreement on which types of crimes are suitable for profiling
        (1) All models agree that rape and sexual murder are suitable for profiling
        (2) Douglas et al (FBI) include threats and hostage-taking
        (3) Douglas et al (FBI) and Turvey include nonsexual murder
        (4) Douglas et al (FBI), Holmes & Holmes, and Turvey include serial murder and arson
        (5) Turvey includes internet crime (though his is the most recent work)
  2. No clear standards to evaluate the contribution of profiling, or its success at various stages of the investigation and trial.

B. **Use of unclear terms and definitions**:
  1. Serial killers defined differently
  2. *Modus operandi* sometimes (Douglas et al (FBI), Turvey) described as aspects of a crime that change, and sometimes (Holmes & Holmes) as aspects that remain the same from crime to crime.
  3. No indication of what a criminal's signature is, or how it should be determined

C. **Misuse of typologies**:
  1. **Inappropriate use of typologies**: Holmes & Holmes, and Keppel & Walter advocate matching offenders to typological categories. This is a problem since offender may meet criteria for several categories, and/or fail to meet all criteria for any one category.
  2. **Inconsistent presentation within typologies**:
     a. Holmes & Holmes:
        (1) make initial distinction between geographically stable vs geographically transient, but make no further use of this categorization, and do not indicate how it interacts with other categories in their scheme.
        (2) provide three categorization scheme for arsonists, one based on organized/disorganized distinction, and two others on the basis of motive. No

information on relationships between these.
- (3) use pedophile and child molester separately in some places, as synonyms in others.
- (4) define power-assertive rapists as 'impulsive', but then note the 20-25 day cycle of their attacks, implying something other than impulsivity.
- (5) note that crime scenes, dumping sites and crime locations are choices by offender, but that is inconsistent with the notion that some offenders are disorganized.
   b. Turvey: Argues that his behavior-motivational typology applies to all crimes, but only mentions sexual behavior of offenders in his category.
   c. Keppel & Walter mention types of statements each type of criminal would make to victim, but this is information that cannot be retrieved if victim is murdered.
3. **Overlap among typological categories**: In most typologies, several different categories include similar or identical characteristics (e.g., young, single male; use of weapons, etc.)
4. **Value of typology**: How does categorizing the offender help the police narrow down the range of suspects? Typological category too much like a horoscope to be useful.

D. **Reliance on intuition and professional knowledge**: How do we validate this if multiple intuitions exist, especially since little done to evaluate profiles as useful to investigation?

E. **Lack of clear procedures**: No indication in any model of how offender characteristics are to be determined from the evidence presented.
1. Douglas et al (1986) model indicates that offender characteristics like height, weight, eye color, hobbies and interests should be included in the final profile, but provides to indication of how one would get this information.
2. Holmes & Holmes (1996) indicate the need to "take into account the total crime scene in order to form a mental image of the personality of the offender." (P. 39) But no indication of how to form this mental image based on the crime scene, or what aspects of the offender's personality are to be represented.
3. For models with typologies, no indication of how to select a type for a given offender when typological categories contain significant overlap.
4. No clear procedures for assessing offender motive and signature, which is important in both Douglas et al and Walter & Keppel model.
5. All nonscientific profiling models that discuss modus operandi (MO) and signature include them as related concepts, and their definitions are not clear.
   a. if these two ideas can be conceptually untangled, there are still no procedures in any of the models for distinguishing and determining them in practice.
6. Some models indicate what sort of evidence to gather as part of the profiling process (Douglas et al, Turvey), while others do not.
7. No clear indication in most models of how to get to the final profile from the evidence.
8. Only two models provide enough procedural info to achieve one stated goal of the model. [see H&S, p 65]
9. Even when procedures provided, they are typically only at the start of the process. ***"For example, both the Douglas et al (1986) model and the Turvey (1999) model offer procedures for collecting evidence and analyzing crime scenes but do not offer***

procedures for reconstructing a crime, linking evidence to offender characteristics, linking offenses, using typologies, or determining MO and signature." (H&S, p. 65)

F. Lack of evidence of investigative value
   1. **Holmes & Holmes** (1996, p. 44) mention an FBI study indicating that only 88 of 192 cases in which profiling was used were solved. Of those 88, profiles identified the offender in only 17% of cases. Suggesting that profiling successful in only 8% of cases in which it was used.
   2. **Gary Copson (1995)**, as cited in **Canter (2000)**, found only 3% of profiles useful.

IX. David Canter's Scientific Model of Profiling

   A. Canter points out that nowhere is there any information about how to produce an offender profile. Even the FBI does not say how they do it, but argues that it is based largely on 'intuition'.
      1. But even though profiles not based on 'intuition' are denigrated by the FBI, in fact most of the information in their profiles are not derived from intuition, but is based on known characteristics of previous offenders.
      2. FBI profiles also contain info about characteristics not possessed by the offender (e.g., no military experience), and these are also based on statistical data, and are often characteristics with a low base rate in the population.
      3. Also criticizes FBI's lack of research, or use of research in profiling.

   B. Canter says profiling inferences and questions come from four categories:
      1. **Behavioral salience**: What are the behavioral features of the crime that may help identify the perpetrator?
      2. **Distinguishing between offenders**: How can we indicate differences between offenders, including differences between crimes?
      3. **Inferring characteristics**: What inferences can we make about an offender that might help us identify him or her?
      4. **Linking offenses**: How can we determine whether a series of crimes were committed by the same perpetrator?

   C. Forensic profiling different from ordinary process of drawing inferences from behavior because:
      1. Information available to forensic profiler is limited to that provided by a crime scene (or scenes.)
         a. Identity of victim
         b. Time and location of the crime
         c. Sequence of events during crime
         d. No direct information available about the psychology of the perpetrator from the evidence of the crime, even if we include the testimony of the victim.

2. Profiler asked to provide only information that will be useful in the police investigation: E.g., physical characteristics, living situation, etc. Information about internal personality dynamics would not be useful.

D. Canter proposes simple general equation to describe the process of linking crime actions and offender characteristics:
1. $F_1A_1 + ... F_nA_n = K_1C_1 + ... K_nC_n$ where each A represents an action of the offender, and each C represents a characteristic of the offender. The problem is deciding on the weights (F or K).
2. Problems are that:
   a. There is no strong relationship between any particular crime action and any specific offender characteristic.
   b. Variations in the actions included in the data set may change the weightings on both sides of the equation: An action included or left out might dramatically change the estimate of one or more offender characteristics.

E. **Interpersonal Narratives**: This is how Canter links crime actions to offender characteristics. He argues that it "attempts to build links between the strengths of all the approaches outlined above " (Canter, 1995, p. 353)
1. A crime is an interpersonal transaction that involves characteristic ways of dealing with other people.
2. Need to identify different interpersonal themes as a first step in identifying offenders.

F. Canter's interpersonal themes that are involved in violent crime: "the crucial distinctions between the dramas that violent men write for themselves are the variations in the roles that they give their victims.... variations in the emphases of the vicious interpersonal contact are therefore the first major themes to consider when interpreting any violent crime. " (Canter, 1994, p. 339) Each of the themes below is orthogonal with the dimension of Desire for Control, which can be either High or Low.
1. **Victim as Object**: Offender has no feeling or connection with victim, who does not play any active role in the assault. Crime likely to be one of opportunity, with victim encountered in nondescript public place.
   a. **High Desire for Control**: Victim may be mutilated with parts being cannibalized or taken away as souvenirs. Conceptually similar to FBI's Disorganized Offender.
   (1) Likely to be of low intelligence, perhaps with lack of contact with reality.
   (2) likely to live alone, or move from institution to institution.
   (3) known in the community as an eccentric.
   (4) comes from dysfunctional background, possible changes of parenting during childhood, perhaps poverty as well.
   (5) crimes likely to come to attention accidentally.
   (6) will be aware that actions are criminal, but may not avoid or resist capture other than by changing location of crimes. If captured, likely to confess.
   b. **Low Desire for Control**: Offender likely to choose victim he finds attractive.
   (1) Sexual component of the crime will be more prominent than acts of mutilation or dismemberment.
   (2) murder a consequence of offender's violent actions during the crime, but not a

major goal of the crime.
- (3) offender obsessed with obtaining more victims. May find a secluded spot where victims can be kept for longer periods of time.
- (4) offender likely to commute to location where attractive victims more likely to be found.
- (5) offender will not have much verbal interaction with victim, and may bring weapons and binding materials to the crime scene.
- (6) offender will be employed in a "non-demanding job" (Canter, 1994, p. 349) that does not involve much contact with people.
- (7) offender will not understand seriousness of his actions, and will respond to questions about crimes with disinterest or nonchalance.

2. **Victim as Vehicle**: Theme here is the offender as tragic hero. He is angry with himself, and with his unsatisfying lot in life. Assaults allow him to regain his rightful place.
   a. **High Desire for Control**: Similar to FBI's category of spree murderer. May kill many people in a single bout. May also intensify the experience by committing suicide after murders, what Canter (1994) calls the "Samson syndrome" (p. 351)
   b. **Low Desire for Control**: Killings more deliberate and serial, as offender knows he has a destructive mission.
   - (1) Would fit FBI description of organized offender. Travels to commit his crimes.
   - (2) motivation stems from some failed or lost relationship - death of loved one, end of intimate relationship. Canter argues that crimes are attempts to rebuild these relationships in his inner narratives.
   - (3) Offender eager to talk to police, to gain recognition, and is eager to have his story presented in the media..
   - (4) offender is intelligent, with good social skills. Uses charm to manipulate victims and gain their trust.
   - (5) will want victim to be exploited.
   - (6) lacks sympathy for victim, or remorse for his actions.

3. **Victim as Person**: Offender sees victim as individual, and tries to understand their experience, though has no true empathy for them. Note that **Canter did not address the control dimension in describing this type of offender.**
   a. May misunderstand feelings behind victim's reactions. E.g., may commit rape, then ask victim for a date. Victims are opportunistic.
   b. Violence is considered normal part of life by the offender.
   c. Offences usually indoors, and assaults are often unplanned sequelae to robbery or home invasion.
   d. Some are teens who attack elderly women known to them as part of burglary or theft. Multiple crimes likely here, since purpose is personal gain.
   e. In rape cases, offender may believe that he is establishing a relationship with the victim, so may stalk her and attack in her home. May initially target women known to him from his neighborhood.

G. Testing Canter's Hypotheses
   1. Used statistical technique called **smallest space analysis** (SSA) to test hypotheses

about interpersonal narratives, offender consistency, and offence specificity. SSA presents related data (e.g. correlations) in a spatial field with their proximity determined by the strength of their relationship.
2. Then used theories to identify dominant themes among variables, and partitioned space to aggregate variables related to different themes.
3. Then either looked for clusters of variables specified in advance, or identified themes by eyeballing the plot.

X. Hicks & Sales provide a lengthy critique of Canter's theory.

  A. Two basic problems with Canter's role themes:
    1. **Lack of conceptual clarity**:
      a. Role of victim in different roles not clearly distinguished.
      b. Inconsistency in descriptions within themes: I.e., says of offender in victim-as-object theme, **BOTH** that he may be of low intellectual ability, **AND** that he may be more intelligent and manipulative.
      c. Describes victim-as-object offenders as lacking contact with reality, having hallucinations which they are unable to distinguish from reality. But argues that they are not psychotic despite the fact that they meet the DMS definitions for schizophrenia.
      d. In victim-as-vehicle, describes offenders as unfeeling psychopaths, but says crime triggered by loss or death of loved one, which would suggest emotional investment uncharacteristic of a psychopath.
      e. Does not mention level of control in describing victim-as-person theme.
      f. Seems to equate 'high' control with physical violence, and 'low' control with psychological manipulation, though both suggest an offender who wants control.
    2. **Unverifiable assertions**: E.g.:
      a. Argues that "sexuality and ... bizarre sexual acts dominate the personal narratives" of victim-as-object offenders. But how would one know this, since personal narratives are intrapsychic.
      b. Notes that victim-as-vehicle offenders use assaults to regain sense of power and freedom that is absent "in the other stories they are forced to live." (Canter, 1994, p. 351). No way to know this.

  B. Operationalization and Selection of Methods & Statistics
    1. Does not define terms so as to indicate how they are to be determined from crime scene evidence, or indicate whether or how it should be a focus of interest in the investigation.
    2. Does not relate concepts to those in other models, even when they seem to be the same. (E.g., MO, signature, interpersonal narrative vs motivation.)
    3. Four main methodological problems with Canter's research: (particularly four studies: **Canter & Ftitzon, 1998; Canter & Heritage, 1990; Canter et al, 1998; Salfati & Canter, 1999**)

      a. **Use of SSA for analysis**: (seems to resemble factor analysis in some ways)
         (1) Does not provide clusters (as factor analysis would), so themes determined by eyeballing data
         (2) Does not link themes with offender characteristics
         (3) Does not indicate predictive power of clusters to identify offender characteristics, which is a prime goal of the model.

C. Methods for profilers in practice: A scientific model must indicate what profilers actually do.. Following points must be addressed in any such description:
  1. **Evidence and information**:
     a. What evidence and information should be collected for profiling purposes?
     b. How does investigator decide what evidence is relevant and important?
     c. How and by whom is evidence gathered?
  2. **Interpretation of evidence**: How is this done? E.g.:
     a. How do we decide whether crime scene is staged or not?
     b. How do we determine the sequence of events?
  3. **Determining offender characteristics**:
     a. What are the relationships between crime scene evidence and offender characteristics?
     b. How do we determine which offender characteristics are important in investigating the crime?
  4. No research linking nature of crime scene evidence to offender characteristics.

XI. Good and Bad profiling

A. Many profiles in prominent cases were well off the mark:
  1. **The D.C. (Beltway) Sniper** case: October 2002, ten people killed by sniper fire in and around the Washington, D.C. area and along Interstate 95..
     a. FBI profiles identified the offender as:
        (1) White male, 25 to 40 years old
        (2) Not a sniper and not likely military
        (3) Lives in or near the community
        (4) No children; Firefighter or construction worker
        (5) Possible terrorist links
     b. John Allen Mohammed: Spent 7 years in Louisiana National Guard (1978-1985), then in U.S. Army from 1985-1994, when discharged after Gulf War with rank of Sergeant. In Army, trained as a mechanic, truck driver and specialist metalworker. Earned "expert" rating with M16, the highest level of marksmanship for a soldier. Was a member of the Nation of Islam for a time; moved to Antigua in 1999, and apparently engaging in credit card and immigration document fraud activities. After his arrest, authorities also claimed that Muhammad admitted that he admired and modeled himself after Osama bin Laden and Al Qaeda, and approved of the

September 11, 2001 attacks. Twice divorced; second wife granted a restraining order. Muhammad was arrested on federal charges of violating the restraining order against him, by possessing a weapon.
2. **Baton Rouge Serial Killer** (2003):
   a. FBI profile: White male, between 25-35, an unsophisticated social outcast who did was awkward around women, and did not get along well with them. Impulsive, with a quick temper and motivated by a desire to retaliate against women. Would brag to friends and co-workers about non-existent relationships. Abusive and distant if in a relationship.
   b. **Derrick Todd Lee**: A religious 34-year old black male with an ex-wife and two kids. Known as smooth, handsome, charming, and outgoing - a 'Casanova' with women, frequenting local bars with a different woman every night. He lived with a girlfriend while separated from his wife.
3. **1996 Summer Olympic bombing** in Atlanta (See Costanzo, *Psychology Applied to Law*, pp. 69-70):
   a. FBI profile: Single, white, middle-class male with an intense interest in police work and law enforcement. Attention focused on Richard Jewell, security guard at Olympic Par who fit the profile.
   b. Eric Rudolph, an anti-abortion activist, charged with the bombing. Aryan superiority believer. Had hatred for U.S. government, but joined military to become an army ranger. Learned about explosives and survival tactics, but discharged for insubordination and marijuana use.

B. But some profiles reasonably accurate:
   1. **Ted Koczynski** (unabomber): Here one of the several profiles was quite accurate:
      a. Several profiles, one of which, by FBI agent Bill Tafoya, was quite accurate:
      b. Initial profile: Male, mid-30s to early 40s, with some college education. Perhaps a blue-collar aviation worker.
         (1) Tafoya's profile: Male, probably in his early 50s (Kaczynski was 53 when captured). Probably college and graduate degree, perhaps even a Ph.D. Background in science, perhaps electrical engineering or math, and a strongly anti-technology.
      c. In fact, Kaczynski entered Harvard as a math whiz in 1958 at the age of 16. After graduation, earned Master's and Ph.D. degrees in mathematics. In 1967, hired as Assistant Professor of mathematics at the Berkeley. Resigned in 1969, and moved first to his parents' home in Lombard, Illinois, then to a remote cabin in Montana.
   2. **Paul Bernardo**: FBI Profile of Scarborough rapist (1988):
      a. Single white male, 18 to 25 years of age. "The behaviour exhibited throughout these assaults suggests a youthful offender rather than an order more mature one." Lives in Scarborough area, probably in the vicinity of the initial assault sites. Angry towards women, and will speak disparagingly of women in conversation with associates. Has had a major problem with women immediately before onset of attacks. Sexually experienced, but past relationships stormy and ended badly. Probably has battered women he has been involved with in past. Places blame for all his failures on women. If he has a criminal record, probably for assault, disturbing the pace, resisting arrest, domestic disturbance, etc.

(1) Aggressive behaviour will have surfaced in adolescence.
(2) High school education with discipline problems. May have received counseling for inability to get along with others, aggressiveness, or substance abuse. Bright, but an underachiever in formal academic settings.
(3) Has explosive temper, easily angered.
(4) Work record sporadic and spotty - cannot hold a job due to inability to handle authority. Financially supported by mother or another dominant female.
(5) Can deal with people on superficial level but prefers to be alone - a lone wolf.
(6) Keeps personal property of victims as trophies to relive the assaults.
(7) Attacks will be sporadic, each precipitated by stress in offender's life.
(8) Has no guilt or remorse; believes anger and attacks are justified. Only concerned about being identified and apprehended.
  b. But interviewed because of resemblance to rape victim's sketch

C. Serial murderers/rapists not caught by profile:
 1. Jeffrey Dahmer: Assaulted man escapes, alerts police.
 2. Ted Bundy: Scuffled with officer after being stopped in a stolen car.

D. Most serial killers/rapists are white (85%) males (88%) in twenties or thirties (mean = 28.5) who target strangers (62%) near home or work (71% operate in specific area.)

XII. Evaluating the success of offender profiling

A. Criticisms of profiling:
 1. Many systems based on theoretical model of personality that lacks strong empirical support (i.e., trait theory; psychodynamic theory)
 2. Assumes dominance of personality over situation in determining micro-aspects of behavior. (Though some note that higher levels of behavioral consistency in pathological cases than in normal)
 3. Many profiles contain information that is so vague and ambiguous they can potentially fit many suspects.
    a. **Alison, Smith, Eastman & Rainbow (2003)**:
    b. **Alison, Smith, & Morgan (2003)**:
    c. See also: **Alison, Bennell, Mokros, & Ormerod (2002)**:
 4. Many aspects of profiles commonsense deductions from evidence
 5. Professional profilers may be no better than untrained individuals at constructing accurate criminal profiles.

B. **Copson (1995)**: Gary Copson is Detective Chief Inspector with London Metropolitan Police in U.K., and reported on 5-year project on profiling supported by the Association of Chief Police Officers and the British Home Office.
 1. Surveyed 184 detectives who had made use of profiling services
 2. Reported that **83%** of police officers surveyed described profiles as operationally

useful.
3. But only **2.7%** indicated that profiles had lead to the identification of the offender.
4. More said that they were useful because they helped them understand the case (**61%**) or because it reassured them of their own judgment about the offender (**51.6%**)
5. In presenting results to British Psychological Association, DCI Copson said: "You may as well toss a coin for some." He added that some profiles "riddled with inaccuracies and inconsistencies which suggest they didn't really do any good."

C. **Kocsis et al (2000)**: Compared accuracy in profiling of 6 groups: 5 profilers (out of 40 invited); 35 Australian police officers; 30 Australian psychologists (3 or more years of training, but not in criminal or forensic psychology; Australian university science and economics 2nd years students; 20 self-declared Australian psychics; 23 Australian economics students as control group.
   1. Given details of solved homicide: Scene of crime report; forensic biologist's report; forensic entomologist's report; ballistics report; preliminary postmortem report; pathologist's postmortem report; basic info on identity/background of victim; schematic plan of crime site; nine photos of crime scene & victim, showing wounds. Same as info available to investigators prior to determination of primary suspect (person ultimately convicted of crime.)
   2. Then profile generation in three stages:
      a. First write detailed description of offender; told it should include any personal and physical characteristic that might help police apprehend the offender.
      b. Then get 45-item m-c questionnaire re offender's physical characteristics, cognitions related to offense; behaviors associated with offense; personal history. Only 33 items relevant to analyses reported.
      c. Finally, complete Adjective Check List (ACL) to describe personality of offender.
      d. Correct answers determined by having officer who worked on case complete the forms.

Measure	Profilers	Psychlgsts	Police	Students	Psychics	Control	Sig
Cognitive processes	3.20	2.27	2.49	2.03	2.60	2.35	
Physical characteristics	3.60	3.63	3.43	3.42	2.80	2.09	**
Offense behaviors	4.00	4.03	3.09	3.64	3.65	3.61	**
Social history, habits	3.00	2.63	2.6	2.94	2.25	1.74	
Personality overall	24.6	34.03	22.03	26.84	27.7	24.48	***
Total	13.8	12.57	11.6	12.03	11.3	9.78	0

* = p<.10; ** = p<.05; ***=p<.01

   e. Results:

(1) Only marginal difference in **overall accuracy** (p<.10). Follow-up measures do not reveal and differences between specific groups, except everybody better than control except psychics, who did not differ.
(2) For **physical characteristics**, psychologists > police; psychologists > psychics. No other significant comparisons.
(3) For **offense behaviors**. Psychologists > police`; students > psychics. No other significant group comparisons.
(4) For **social history/habits**: No group differences.
(5) For **personality** (ACL): psychologists > police, but confounded with number of adjectives checked on ACL. When this factor controlled, no difference between groups.
(6) Correct answers determined by having officer who worked on case complete the forms.

XIII. Profile of Jack the Ripper by John E. Douglas and Roy Hazelwood

A. Done for two-hour October 1988 broadcast, "The Secret Identity of Jack the Ripper", hosted by Peter Ustinov. The 100th anniversary of the Whitechapel murders

B. "As in the Yorkshire Ripper case 90 years later, we were convinced the taunting letters to the police were written by an imposter, someone other than the 'real' Jack. The type of individual who committed these crimes would not have the personality to set up a public challenge to the police. The mutilation suggested a mentally disturbed, sexually inadequate person with a lot of generalized rage against women. The blitz style of attack in each case also told us he was personally and socially inadequate. This was not someone who could hold his own verbally. The physical circumstances of the crimes told us that this was someone who could blend in with his surroundings and not cause suspicion or fear on the part of the prostitutes. He would be a quiet loner, not a macho butcher, who would prowl the streets nightly and return to the scenes of his crimes. Undoubtedly the police would have interviewed him in their investigation. Of all the possibilities we were presented, [Aaron] Kosminski fit the profile far better than any of the others. As for the supposed medical knowledge needed for the post-mortem mutilation and dissection, this was really nothing but elementary butchery." (Douglas, MindHunter, pp 174-175.)

References

Alison, L.J., Bennell, C., Mokros, A. & Ormerod, D. The personality paradox in offender profiling: A theoretical review of the processes involved in deriving background characteristics from crime scene actions. *Psychology, Public Policy, and Law*, 2002, 8, 115-135.

Alison, L.J., Smith, M., & Morgan, K. Interpreting the accuracy of offender profiles. *Psychology, Crime, and Law*, 2003, 9, 185-195.

Alison, L.J., Smith, M., Eastman O. & Rainbow, L. Toulmin's philosophy of argument and its relevance to offender profiling. *Psychology, Crime, and Law*, 2003, 9, 173-183.

Arrigo, B. A., & Shipley, S. L. (2005). Introduction to Forensic Psychology: Issues and Controversies in *Law, Law Enforcement and Corrections* (2nd ed.). New York: Elsevier.

Brussel, J. *Casebook of a Crime Psychiatrist*. New York, Bernard Geis Associates, 1968.

Canter, David. *Criminal Shadows*. London, HarperCollins, 1988/1994.

Canter, David. (1995). Psychology of offender profiling. In R. Bull & D. Carson (Eds.), *Handbook of Psychology in Legal Contexts*. New York, Wiley. [$250 !!]

Canter, David. (2000). Offender profiling and criminal differentiation. *Legal and Criminological Psychology*, 5, 23-46.

Canter, David. (2004). Offender profiling and investigative psychology. *Journal of Investigative Psychology and Offender Profiling*, 1, 1-15.

Canter, D. & Fritzon, K. (1998). Differentiating arsonists: A model of firesetting actions and characteristics. *Journal of Criminal and Legal Psychology*, 3, 73-96. [Journal may no longer exist. May be earlier version of Legal and Criminological Psychology before 1999.]

Canter, D. & Heritage, R (1990). A multivariate model of sexual offense behavior: Developments in 'offender profiling'. *Journal of Forensic Psychiatry*, 1, 185-212. [not available online]

Canter, D. Hughes, D. & Kirby, S. (1998). Paedophilia: pathology, criminality, or both? The development of a multivariate model of offense behavior in child sexual abuse. *Journal of Forensic Psychiatry*, 9, 532-555. [not available online]

Canter, D.V. & Wentink, N. (2004) An empirical test of Holmes and Holmes's serial murder typology. *Criminal Justice and Behavior*, 31:489–515. ^pdf

Canter DV, Fritzon K (1998) Differentiating arsonists: a model of firesetting actions and characteristics. *Legal and Criminological Psychology*, 3:73–96. Pdf

Copson, G. (1995). Coals to Newcastle? Part 1: A study of offender profiling (paper 7). London: Police Research Group Special Interest Series, Home Office.

Doan, B., & Snook, B. (2008). A failure to find empirical support for the homology assumption in criminal profiling. Journal of Police and Criminal Psychology, 23, 61–70.

Douglas, J.E., Burgess, A., Burgess, A. & Ressler, R. **Crime Classification Manual**. New York, Lexington Books, 1992.

Douglas, J.E., Ressler, R., Burgess, A., & Hartman, C. Criminal Profiling from Crime Scene analysis. *Behavioral Sciences and the Law*, 1986, Vol. 3, pp. 401-421.

Groth, A., Burgess, A., & Holmstrom, L. Rape: Power, Anger, and Sexuality. *American Journal of Psychiatry*, 1977, 134, 1239-1243.

Hazelwood, R.R. & Burgess, A.N. Practical Aspects of Rape Investigation: A Multidisciplinary Approach. New York: Elsevier North-Holland, 1987.

Hicks, Scotia J. & Sales, Bruce D. (2006). *Criminal Profiling: Developing an Effective Science & Practice*. American Psychological Association.

Holmes, R. & DeBurger, J. Profiles in Terror: The Serial Murderer. *Federal Probation*, 1985, 49(3), pp. 29-34.

Holmes, Ronald., & Holmes, Stephen. Profiling Violent Crimes: An Investigative tool. Thousand Oaks, CA, Sage Publications, 1996.

Keppel, R. & Walter, R. Profiling killers: A revised classification model for understanding sexual murder. *International Journal of Offender Therapy and Comparative Criminology*. 1999, 43, 417-434.

Kocsis, R.N., Irwin, H.J., Hayes, A.F. & Nunn, R. Expertise in psychological profiling: A comparative assessment. *Journal of Interpersonal Violence*, 2000, 15-311-331.

Liebert, J. Contributions of psychiatric consultation in the investigation of serial murder. *International Journal of Offender Therapy and Comparative Criminology.* 1985, 29, 187-188.

Mokros, A., & Alison, L. J. (2002). Is offender profiling possible? Testing the predicted homology of crime scene actions and background characteristics in a sample of rapists. *Legal and Criminological Psychology*, 7, 25–44.

Ressler, R.K, Burgess, A., & Douglas, J. Sexual Homicide: Patterns and Motives. New York, Lexington Books, 1988.

Rider, A. The Firesetter: A Psychological Profile (Part 1). *FBI Law Enforcement Bulletin*, 1980a, 49(6), 6-13.

Rider, A. The Firesetter: A Psychological Profile (Conclusion). *FBI Law Enforcement Bulletin*, 1980a, 49(6), 12-17. [Available online only from 1996]

Rumbelow, Donald. Jack the Ripper: The Complete Casebook. Contemporary Books, 1988.

Salfati, C.G. & Canter, D. (1999). Differentiating stranger murders: Profiling offender characteristics from behavioral styles. *Journal of Behavioral Sciences & the Law*, 17, 391-406.

Turco, R. Psychological Profiling. *International Journal of Offender Therapy and Comparative Criminology*, 1990, 34, 147-154.

Turvey, B. Criminal Profiling: An Introduction to Behavioral Evidence Analysis. San Diego, CA, Academic Press, 1990.

Jury Psychology

I. Jury Selection

    A. Much of jury selection research based on fact that attorneys have multiple challenges too prospective jurors from the jury pool:
        1. Challenges for cause:
        2. Peremptory challenges:
            a. In Canadian criminal law, under section 634, both Crown and defense entitled to:
                (1) 20 peremptory challenges, in cases of treason or first degree murder
                (2) 12 peremptory challenges in cases of offence that carry a penalty of more than 5 years in prison
                (3) 4 peremptory challenges otherwise
                (4) But in Canada, lawyers not allowed to question porential jurors extensively, and only know the candidate's name, area of residence, and profession.
            b. In U.S.
                (1) 20 peremptory challenges when the government seeks death penalty.
                (2) Government has 6 peremptory challenges and defendant has 10 peremptory challenges when defendant is charged with a crime punishable by imprisonment of more than one year.
                (3) Each side has 3 peremptory challenges when defendant is charged with a crime punishable by fine, imprisonment of one year or less, or both.
                (4) In U.S., extensive questioning of jurors is common, though rejection on the basis or race or gender is forbidden by Supreme Court decision.

    B. History:
        1. Began with politically-oriented trials in early 1970s, esp 1972 trial of Harrisburg Seven.
            a. Group (Phillip and Daniel Berrigan) charged by FBI with conspiring to destroy draft board records, kidnap Henry Kissinger, destroy heating tunnels in D.C.
            b. Government chose to hold trial in conservative city of Harrisburg, PA
            c. Sociologist Jay Schulman and social scientists helped defence in jury selection
            d. Surveys to determine characteristics related to verdict preference to be asked during *voir dire*.
                (1) phone interviews with 840 registered voters; 250 detailed in-person interviews
                (2) Asked about:
                    (a) media exposure to trial info; knowledge re defendants; trust in government; religious attitudes; ages/activities of children;
                    (b) specific trial-related attitudes:
                        i) "People should support their country even when they feel strongly that the federal authorities are wrong."
                        ii) "If the authorities go to the trouble of bringing someone to trial in a court, the person is almost always guilty."
                    (c) Anti-war attitudes on a scale:
                        i) "Accept what the government is doing and keep quiet about one's feelings."

    ii) "Become part of a revolutionary group that attempts to stop the government from carrying on the war by bombing buildings or kidnapping officials."
  e. Result was hung jury with 10 for acquittal. Defendants never retried. Two who voted for conviction would have been eliminated by challenge, but one lied about his beliefs, and the other was accepted because she indicated that her sons were conscientious objectors. Further questioning would have revealed that she strongly disagreed with heir stance.
2. Schulman and team (including psychologist Bruce Sales) also helped with several other politically-charged trials in early 1970s, then founded National Jury Project (**www.njp.com**), the first firm to do professional jury consulting.
3. Jury selection now much more common in civil trials where damage awards involved
4. Donald Vinson founds Litigation Sciences in 1979 sold in 1987), and then Decision/quest (**www.decisionquest.com**) the firm that provided jury selection expertise in the O.J. Simpson trial. Website lists advantages of jury research in civil trials:
  a. **Discloses the questions jurors want asked in discovery**: "Conducting exploratory research beforehand, allows you to construct a line of questioning that can reveal both the good and bad points about a witness. [It] also allows group of individuals similar to actual jury to identify deposition questions they want asked."
  b. **Identifies problems in the case early, so you will have time to fix them**: "One of the main benefits of doing jury research early is that, by finding blind spots, you have time to prepare answers to questions you would not otherwise expect to be asked. Just as important, early research may also disclose the opponents' own blind spots."
  c. **Removes your blinders in time to see the finish line**: "jurors convey to you what worked, and what did not work, after they have heard both sides of the case and have begun to develop their initial perceptions. A lucid understanding of how the jurors perceive the facts of the case will provide guidance for how to best map your strategies and prepare your arguments."
  d. **Arms your side for mediation**: "Opposing counsel may be unsettled if they sense that you have gained knowledge that could probably have been gained only through research. The research could thus force their hand for two reasons: because of what it uncovers, and because they know it is an ongoing tool in your possession."
  e. **Provides a range of realistic damage values**: "a trial simulation ... allows counsel to be more confident in their approach once they get a range of numbers from informed jurors who have heard a compact version of both sides of the case."
  f. **Provides guidance on the graphics to be used**: "jury research discloses the information that jurors want to better understand your case. In this context, they will specifically ask for graphics to simplify complicated aspects of the case. It is true that different people learn more or less visually. It has become readily apparent that well-constructed graphics are key to winning almost any type of case."

C. O.J. Simpson trial (1995)
1. Defense team had jury-selection assistance from Jo-Ellen Demetrius, (who also helped with jury selection in the Rodney King beating case).
  a. She conducted extensive pre-trial research including community surveys, focus groups, and had identified pretrial opinions of people in L.A. with backgrounds

                similar to those of potential jurors.
            b. Determined that black women had overwhelming support for O.J., and disliked prosecutor Marcia Clark.
            c. Listed best candidates for defense as: Young, blue-collar, less well-educated, low SES, Black women.
            d. Defense used challenges to rid jury of any who did not match this profile.
            e. Final jury consisted of 1 Black male, 8 Black women, 2 White women, 1 Latino male.
        2. Donald Vinson offered his services *pro bono* to the prosecution, but they dismissed him after a few days. His research had suggested the same thing.
        3. Methods:
            a. **Mock juries**: presented with trial evidence before the trial, and their judgments noted.
            b. **Shadow juries**: Sit in the courtroom during the trial and brief their clients at the end of the day.
            c. **Community surveys**: To identify undesirable jurors.
            d. **Focus groups**:

D. Juror demographic characteristics and criminal verdicts:
    1. **Occupation**: No indication of relevance.
    2. **Socioeconomic status** (income, education): Results vary considerably. Some find that higher SES associated with higher conviction rate (e.g., **Adler, 1973**), others find that high SES associated with higher levels of acquittal (e.g., Moran & Comfort, 1982). Others (Simpson, 1967) find no relation at all.
    3. **Education**: Again, some find higher conviction rates with higher education (e.g., Reed, 1965), some find lower (e.g., Weiner & Stolle, 1997); others find no relation (e.g., Moran & Comfort, 1982)
    4. **Age**: Another wash-out. Some studies show higher conviction rates for older (e.g., Wiener & Stolle, 1997), others find no relationship (e.g., Moran & Comfort, 1982). May depend on crime; Mills & Bohanon (1980a) found older jurors more conviction prone in rape, cases, but less so in murder cases.
    5. **Gender**: No clear relationship.
    6. **Race-ethnicity**: Mixed bag again.
    7. **Religion**: Both real jurors and mock jurors show no relation between religion and verdict.
    8. Though no clear overall effects (which is a good thing), results for specific cases may be quite clear - as in the O.J. Simpson case.

E. Juror personality factors and verdicts
    1. **Authoritarianism**: Desire for order, well-defined rules, authoritative leadership
    2. Several tests to measure it:
        a. **California F-scale** (See http://www.anesi.com/fscale.htm)
            (1) "Obedience and respect for authority are the most important virtues children should learn."
            (2) "The business man and the manufacturer are much more important to society than the artist and the professor."
            (3) "Every person should have complete faith in some supernatural power whose

decisions he obeys without question."
- (4) "What this country needs most, more than laws and political programs, is a few courageous, tireless, devoted leaders in whom the people can put their faith."
- (5) "What the youth needs most is strict discipline, rugged determination, and the will to work and fight for family and country."
- (6) "An insult to our honor should always be punished."
- (7) "Sex crimes, such as rape and attacks on children, deserve more than mere imprisonment; such criminals ought to be publicly whipped, or worse."
- (8) "Most of our social problems would be solved if we could somehow get rid of the immoral, crooked, and feebleminded people."
- (9) "Homosexuals are hardly better than criminals and ought to be severely punished."
- (10) "People can be divided into two distinct classes: the weak and the strong."
- (11) "No weakness or difficulty can hold us back if we have enough will power."

b. **Altemeyer Right-Wing Authoritarianism Scale** - RWA:
- (1) Defined as co-existence of three attitudinal clusters:
  - (a) **Authoritarian submission**: Deference and submission to established, legitimate authorities
  - (b) **Authoritarian aggression:** An aggressiveness against persons perceived to be sanctioned by established authorities.
  - (c) **Conventionalism:** Adherence to conventions endorsed by society and established authorities.
- (2) 30 Items on the scale:
  - (a) "Our country desperately needs a mighty leader who will do what has to be done to destroy the radical new ways and sinfulness that are ruining us.
  - (b) "The real keys to the 'good life' are obedience, discipline, and sticking to the straight and narrow,"
  - (c) "Some of the worst people in our country are those who do not respect our flag, leaders, and the normal ways things are supposed to be done."
  - (d) "There is nothing immoral or sick in somebody's being a homosexual"
  - (e) "It is important to protect fully the rights of radicals and deviants"
  - (f) "The way things are going in this country, it's going to take a lot of 'strong medicine' to straighten out the troublemakers, criminals, and perverts")
  - (g) "It is wonderful that young people today have greater freedom to protest against things they don't like and to 'do their own thing' "
  - (h) "Everyone should have their own lifestyle, religious beliefs, and sexual preferences, even if it makes them different from everyone else."
- (3) Right-wing does not refer to political views; In Soviet Union, high RWA persons supported the Communist Party

3. **Authoritarians**:
   a. More ready to convict: E.g., Lamberth, Krieger, & Shay, 1982; Moran & Comfort, 1982)
   b. Have better recall for prosecution than for defense evidence: Garcia & Griffiit (1978)
   c. Recommend longer sentences for convicted defendants: Bray & Noble (1978); McGowen & King (1982)
   d. More punitive towards defendants of low status: Berg & Vidmar (1975); but perhaps

less punitive towards defendant who is authority figure of who has committed a crime by obeying orders (Hamilton, 1978; Garcia & Griffitt, 1978) or who committed murder while disciplining a disobedient son.
4. But meta-analysis by **Narby et al (1993)** found relation between authoritarianism and verdicts to be very small. Effect stronger among actual jurors than among student mock jurors.
5. **Juror Bias Scale (Kassin & Wrightsman, 1983)**
   a. Ps rate 22 statements on 5-point scale (1=Strongly Agree; 5=Strongly Disagree) with 22 statements divided into two sub-scales:
      (1) **Probability of Commission**:
         (a) "A suspect who runs from the police most probably committed the crime."
         (b) "In most cases where the accused presents a strong defense, it is only because of a good lawyer."
         (c) "Out of every 100 people brought to trial, at least 75 are guilty of the crime with which they are charged"
         (d) "Generally, the police make an arrest only when they are sure about who committed the crime."
      (2) **Reasonable Doubt**.
         (a) "If a majority of evidence - but not all of it - suggests that the defendant committed the crime, then the jury should vote not guilty."
         (b) "Circumstantial evidence is too weak to use in court"
   b. Authors report correlations of .31 between verdicts and scale scores in mock trials.
   c. Because of high correlations (~.60) between sub-scales, not clear what JBS measures (See **Penrod & Cutler, 1987**)
   d. **Myers & Lecci (1998)** suggest that Probability of Commission scale itself composed of two factors: Confidence in the Legal system, and Cynicism Toward some aspect of the criminal justice system.
      (1) **Confidence** is three items:
         (a) "A suspect who runs from the police most probably committed the crime"
         (b) "Generally, the police make an arrest only when they are sure about who committed the crime."
         (c) "If a grand jury recommends that a person be brought to trial, then that person probably committed the crime"
      (2) **Cynicism** also in three items:
         (a) "In most cases where the accused presents a strong defense, it is only because of a good lawyer."
         (b) "Defense lawyers don't really care about guilty or innocence, they are just in the business to make money"
         (c) "Many accident claims filed against insurance companies are phony"
      (3) Three items on Probability of Commission scale have little predictive power:
         (a) "Out of every 100 people brought to trial, at least 75 are guilty of the crime with which they are charged."
         (b) "Circumstantial evidence is too weak to use in court."
         (c) "The defendant is often a victim of his own bad reputation"
      (4) Two items on Reasonable Doubt scale were poor predictors:
         (a) "The death penalty is cruel and inhumane."
         (b) "Too many innocent people are wrongfully imprisoned.

(5) Still, revised JBS accounted for only 3.5% of variance in verdicts, compared with 2.6% for the original version.
6. **Dogmatism**: A more general form of authoritarianism characterized by inflexibility and close-mindedness. Not the same connection with prejudice as authoritarianism. Results similar to those for authoritarianism:
    a. High in dogmatism -> more likely to convict, and more likely to be punitive. (E.g., Shaffer & Wheatman, 2000; Shaffer et al, 1986). But not all studies find this, as usual.
7. **Locus of Control**: (Rotter, 1966: Internal-External Locus of Control Scale). Those with external locus of control believe that situational factors outside the individual's control are largely responsible for events, whereas those with internal locus of control believe they control their own destiny. Scale contains paired statements, and Ps choose the one closest to their own beliefs:
    a. E.g.: "As far as world affairs are concerned, most of us are the victims of forces we can neither understand nor control" vs "By taking an active part in political and social affairs, the people can control world events."
    b. "Many of the unhappy things in people's lives are due to bad luck" vs "People's misfortunes arise from the mistakes they make."
    c. **Sosi (1974)**: Found that in drunk driving case, internals recommend harsher punishments; viewed defendant as more responsible. But no relation between locus of control and perceived guilt.
    d. **Phares & Wilson (1972)**: Found that in automobile accident case, Internals attribute more responsibility to defendants when their actions were ambiguous and injury to others was high. No difference when defendant clearly at fault, or when injuries were not severe.
8. **Just World Beliefs**: (cf Montada & Lerner, 1998) involves belief that people generally get what they deserve. In U.K. survey 25 years ago, **Wagstaff (1982)** found that 33% of respondents agreed that victims of rape are usually responsible for the attack.
    a. Measured with a number of instruments, including **Rubin & Peplau's (1975)** Just World Scale. Ps indicate level of agreement with items, e.g.:
        (1) "By and large, people deserve what they get."
        (2) Careful drivers are just as likely to get hurt in accidents as careless ones."
        (3) "Basically, the world is a just place."
        (4) "Crime doesn't pay."
        (5) "People who meet with misfortune often have brought it on themselves."
        (6) "It is rare for an innocent person to be wrongly sent to jail."
    b. **Gerbasi, Zuckerman, & Reis (1977)**: High JW believers in real and mock homicide cases had less favorable impression of defendant, and recommend harsher sentences. Effect seems stronger in women.
    c. **Zuckerman & Gerbassi (1975)**: Rape victim held more responsible by individuals high in JW beliefs. Victims of high moral character held as less responsible than individuals of questionable moral character.

II. Pretrial Publicity

    A. Research in U.S. about extent and nature of pretrial publicity (PTP):
        1. (E.g., **Imrich et al, 1995**) suggests that defendants described in a prejudicial manner in 27% of the stories they examined (14 large U.S. newspapers over 8 weeks)
            a. Mostly about subject's character, and opinions of guilt
            b. Then about prior arrest/conviction information
            c. Most common source of such information was law enforcement officers and prosecutors

    B. Studies can be field surveys of the potential juror pool in actual cases, though few of these have been published, and they generally do not indicate whether the effect of PTP (largely sampled before the trial) persists through the presentation of evidence.

    C. Experimental studies are lower in ecological validity but give us more detailed information about the effects of PTP.
        1. Early study by **Tans & Chafee (1966)**: Varied kinds of PTP given to participants (e.g., seriousness of crime, favorable/unfavorable statements by DA, confession or denial by suspect, was defendant released, or kept in custody). Defendant most likely to be judged guilty when all elements present and unfavorable.
        2. Similar results from subsequent studies:
            a. **Hvistendahl (1979)**:
            b. **DeLuca (1979)**
        3. **Kramer, Kerr & Carroll (1990)**: Compared factual vs emotional PTP, and found 20% higher conviction rate with emotional than with factual. Analysis of deliberations suggested that PTP increased persuasiveness of jurors who argued to convict. Judicial instructions did not reduce the effects.
            a. "Two different types of pretrial publicity were examined: (a) factual publicity (which contained incriminating information about the defendant) and (b) emotional publicity (which contained no explicitly incriminating information, but did contain information likely to arouse negative emotions).
            b. "Neither instructions nor deliberation reduced the impact of either form of publicity; in fact, deliberation strengthened publicity biases. Both social decision scheme analysis and a content analysis of deliberation suggested that prejudicial publicity increases the persuasiveness and/or lessens the persuasibility of advocates of conviction relative to advocates of acquittal
            c. "A continuance of several days between exposure to the publicity and viewing the trial served as an effective remedy for the factual publicity, but not for the emotional publicity."
                (1) Found that:
                    (a) juries exposed to pretrial publicity more likely to render guilty verdict than juries without such exposure.
                    (b) juries exposed to pretrial publicity with high factual content more likely to render guilty verdict when there was no delay between receiving information and the trial.
                    (c) juries exposed to publicity with low factual content more likely to render guilty verdict when there was a delay between publicity and trial.

4. **Penrod & Otto (1992)**: Student jurors register verdicts before and after viewing edited video of actual trial. Verdicts significantly affected by PTP even after viewing evidence. Strongest effects due to negative PTP re defendant's character.
5. **Otto et al (1994)**: Did same as above with actual case involving racial incident involving two campus fraternities. "There were relationships among measures of PTP recall, bias in recall, pretrial attitud4es (e.g., attitudes about fraternities), and ratings of the defendant's culpability." (P. 262)
6. **Steblay et al (1999)**: Meta-analysis of 23 PTP studies.
    a. Participants exposed to negative PTP more likely to judge defendants guilty compared to Ps exposed to less or no PTP.
    b. Larger effects in studies using members of potential jury pool as Ps, rather than students.
    c. Larger effects in studies using real PTP; or involving murder, sexual abuse, or drugs.
    d. Larger effects in studies wither longer delay between PTP exposure and verdict time.
7. **Ruva & Guenther (2015)** one week before viewing criminal trial, 320 mock-jurors (university students) received negative pre-trial publicity (Neg-PTP) or unrelated crime stories (No-PTP). Two days later deliberating jurors came to a decision, while non-deliberating jurors completed unrelated task before making individual decision.
    a. Neg-PTP jurors more likely to:
        (1) Vote guilty
        (2) Make memory errors
        (3) Rate defendant lower in credibility.
    b. Deliberation reduced Neg-PTP jurors' memory accuracy and No-PTP jurors' guilty verdicts (leniency bias).
    c. Jurors' memory and ratings of defendant and prosecuting attorney mediated the effect of PTP on guilt ratings.
    d. Second part of study analyzed content of 30 mock-jury deliberations and explored how PTP influenced deliberations and ultimately jury decisions.
        (1) Neg-PTP juries, compared with No-PTP juries:
            (a) more likely to discuss ambiguous trial evidence in a proprosecution manner
            (b) less likely to discuss judicial instructions and lack of evidence.
        (2) All Neg-PTP juries mentioned PTP, after instructed otherwise, and rarely corrected jury members who mentioned PTP.
        (3) Discussion of ambiguous trial evidence in a proprosecution manner and lack of evidence significantly mediated the effect of PTP on jury-level guilt ratings.
        (4) Findings suggest that judicial admonishments and deliberations may not reduce PTP bias.

D. Effectiveness of legal safeguards
1. *Sub judice* rule: In Canada and many Commonwealth countries, judge can block publication of prejudicial PTP.

III. Jury Decision Making

   A. Models of Jury Decision-Making
      1. Juries spend about 70-75% of time discussing evidence, and 20% of time discussing the law
      2. **Stasser (1992)**; **Costanzo (2003)**: Jury typically moves through three stages during deliberation:
         a. **Orientation**: Elect foreman, discuss procedures, raise general trial issues. Two major deliberation styles (**Hastie, Penrod & Pennington, 1983**):
            (1) **verdict-driven**: About 30% of juries; start with straw poll, and then sort evidence around the results of the poll. Advocate one position at a time, and take frequent straw polls. Reach verdicts faster than evidence-drive juries, but may not look at all the evidence.
            (2) **evidence-driven**: Begin by focusing on evidence. Look at different verdict categories until evidence has been thoroughly reviewed. Individual jurors may relate evidence to several verdicts at the same time. Once jurors agree on best overall account of the evidence, take a straw poll. Generally more rigorous than verdict-driven juries in their examination of evidence.
         b. **Open conflict**: Members attempt to sway each other.
            (1) *normative influence*: Jurors maintain their private beliefs, but change their vote to go along with the others. Most likely when making subjective decisions, such as amount of damages to be awarded.
            (2) *informational influence*: Jurors internalize the arguments on the basis of the evidence, and change their minds and their verdicts. Most likely when making fact-based decisions, e.g. liability, culpability.
         c. **Reconciliation**: Attempts to make sure that everyone is satisfied with the verdict.
      3. Best predictor of a jury's final verdict is the initial verdicts of the individual jurors (**Kalvin & Zeisel, 1966**). Majority generally sways the minority. Discussion tends to strengthen the initial positions - "group polarization".
      4. **Devine et al (2001)**: In criminal trials, when fewer than 8 jurors initially favor conviction, jury is likely to acquit. If 10 or more favor conviction, conviction likely. For 8 or 9 jurors initially favoring conviction, results hard to predict. See **Kerr et al (1996)**; **MacCoun & Kerr (1988)** re *leniency bias*.

   B. **Mathematical Models** of Juror Decision-Making
      1. **Bayesian probability model**: The dominant mathematical model. See **Hastie (1993)** Jurors keep an inner "mental meter" reflecting their belief in defendant's guilt or innocence. Jurors start with initial view about guilt, then assign each bit of evidence a truth probability value, and combining that value with their existing belief in a Bayesian way. Meter stops if probability reaches 1 or 0. At the end of the trial, jurors check whether inner meter has passed their threshold for deciding on guilt.
         a. Not much empirical support for this model. Studies generally find that jurors underuse probabilistic evidence compared with Bayesian norms (See Schklar & Diamond, 1999; Smith et al, 1996)
      2. **Algebraic model**: Similar to Bayesian, but meter does not freeze unchangeably at 1 or 0, and combination of evidence to existing belief is additive rather than multiplicative. Each bit of

evidence assigned a weight. Some empirical support (Moore & Gump, 1995). Model implies that attorneys should present only their strongest evidence.
3. **Stochastic process model**: Similar to above. Begins with juror's initial opinion, and changes to mental meter as each piece of evidence comes in. Mentions 'critical event', which may cause meter to freeze where it is, and no additional evidence makes any difference. At end of evidence, juror's determine whether meter level is above their threshold for conviction. Confidence in verdict depends on meter's distance above (or below) conviction threshold. Some empirical support: Corr & Werner (1994); model successfully predicted changes in juror's assessments of guilt at several point during a trial.
4. Critiques of mathematical models: **Ellsworth & Mauro (1998)**; **Pennington & Hazstie (1981)**:
    a. Models use single meter, but law implies that decision is multi-dimensional, asking jurors to determine whether defendant committed each element of the crime beyond reasonable doubt.

C. **Explanation-based or Cognitive Models** of Juror Decision-Making: The Story Model (**Pennington & Hastie, 1986**)
   1. Juror integrates evidence into story, or causal chain of events, using:
       a. evidence presented during the trial
       b. personal knowledge about similar events
       c. expectations about what a complete story should look like
   2. Since trial testimony is not in story form (out of order, incomplete, etc.) juror must actively create the whole story, including missing links and actor motivations.
   3. Three aspects of story contribute to its credibility and acceptability:
       a. **Coverage**: Does it account for all the evidence presented?
       b. **Coherence**:
           (1) *Consistency*: Is not internally inconsistent, or inconsistent with any of the pieces of evidence.
           (2) *Plausibility*: Is the story similar to the juror;'s understanding of what typically happens in situations like this?
           (3) *Completeness*: Does the story have all the pieces that the juror sees as part of a complete story?
       c. **Uniqueness**: Is this the only story that explains the trial evidence? If so, confidence in it increases. If not, confidence decreases.
   4. Jurors then compare their stories against the various verdict options (which they represent internally as categories defined by lists of features) to reach a verdict.
   5. Considerable empirical support:
       a. **Pennington & Hastie (1992)**: Lab study.
       b. **Olsen-Fulero & Fulero (1997)**: Rape cases.
       c. **Huntley & Costanzo (2003)**: Sexual harassment cases.

D. Jurors' Evaluation of Evidence
 1. **Eyewitness testimony**: **Cutler, Penrod & Struve (1988)**: Only eyewitness confidence affects the impact of testimony on the jurors. Overall, not very good at evaluating such evidence.
    a. Secondary confessions: When defendant confesses to a third party, and that person reports the confession to the police.
    b. **Wetmore, Neuschatz & Gronlund (2014)** with mock jurors, founf that the prsence of a secondary confession led to higher rates of conviction.
 2. **Hearsay evidence**: Jurors good at discounting hearsay evidence (e.g., **Rakos & Landsman, 1992**; **Miene et al, 1992**)
 3. **Confessions**: Jurors appear to assign much higher weight to confession evidence than to any other kind (**Kassin & Neumann, 1997**). This is true even when confession was obtained under coercive means (**Kassin & Sukel, 1997**). Effect found even when confession ruled inadmissible and jurors report that it did not influence their decision.
 4. **Complex evidence**: Too much evidence reduces jurors' self-described understanding of the case, and their confidence in their judgment.
 5. **Expert evidence**: Jurors are generally swayed by expert evidence, ignoring information about construct validity (see **Kovera, McAuliff & Hebert, 1999**)

E. Extra-Evidentiary Factors Affecting Juror Decisions
 1. **Defendant attractiveness**: See meta-analytic study by **Mazzella & Feingold (1994)**.
    a. Mock jurors less likely to find attractive defendants guilty.
    b. If convicted, attractive defendants likely to get lower penalties, but depends on type of crime:
       (1) jurors more lenient on attractive defendants for robbery, rape and cheating.
       (2) jurors harsher on attractive defendants for crimes such as negligent homicide, swindling.
    c. Some evidence (**Bull & Rumsey, 1988**) that unattractiveness is related to criminality. But then there's Jeffrey Dahmer and Ted Bundy...
 2. **Defendant race**: See meta-analysis by **Mazzella & Feingold (1994)**. No evidence of effect on verdicts, but if convicted:
    a. Blacks receive harsher punishments for negligent homicide
    b. White receive harsher punishments for fraud and embezzlement
 3. Studies of race and attractiveness study each separately. **Maeder, Yamamoto & Saliba (2015)** varied both in an alleged acquaintance sexual assault trial. Research done at Carleton University. Mock jurors read trial transcript in which defendant race and victim physical attractiveness manipulated via photos.
    a. Female jurors not influenced by victim attractiveness
    b. Male jurors more certain of defendant's guilt when victim unattractive.
    c. Defendant race and victim attractiveness interacted re victim responsibility
       (1) When defendant White, attractive victims rated more responsible for alleged assault than unattractive victims
       (2) when defendant Black, unattractive victims rated more responsible for alleged assault than attractive victims
       (3) No interaction with Aboriginal Canadian defendants.
 4. **Defendant SES**: See meta-analysis by **Mazzella & Feingold (1994)**. More guilty verdicts and harsher punishments for low-SES defendants, but effect sizes small, and not all studies found

the effect. Two studies actually found harsher punishments for high-SES defendants than for low.
5. **Defendant/juror interactions**: May depend on relationship/similarity of defendant to juror. Some studies find that jurors render fewer guilty verdicts when defendants resemble them in ethnicity, background, or beliefs. (**Amato, 1979; Stephan & Stephan, 1986; Kerr et al, 1995**). Interestingly, research suggests that in cases where racial issues are salient, white jurors appear unbiased, but in cases where racial issues are not salient, they judge black defendants as more violent, aggressive, and guilty. Black jurors more inclined to be lenient to black defendants. (**Sommers & Ellsworth, 2000**)
6. **Juror emotions: Nunez et al (2015)** examined effects of anger, fear, and sadness on jurors' sentencing decisions and explored whether the cognitive appraisal theory or the intuitive prosecutor model could explain these effects. Jurors viewed sentencing phase of a capital murder trial and asked to sentence the defendant.
    a. After viewing, jurors reported increased anger and increased sadness (no increase in fear).
    b. Jurors reporting greater change in anger more likely to sentence defendant to death. No effect of increased sadness.
    c. Effect mediated by importance jurors placed on prosecution's evidence and argument: Increased anger = higher importance rating of aggravating evidence, and more death sentences.

F. Inadmissible evidence:
  1. Can come from pretrial publicity, though less commonly in Canada where judges can order publication bans.
  2. Research indicates that judges instructions to ignore inadmissible evidence are not effective in mitigating the effects of such evidence on juror decisions. (See **Saks & Wissler, 1984; Kasin & Sukel, 1997; Sue et al, 1973**)
  3. But other studies (**Elliot et al, 1988; Kerwin & Shaffer, 1994; Landsman & Rakos, 1994**) indicate that proper judicial instructions can mitigate the effects of inadmissible evidence.
  4. One form of inadmissible evidence is a defendant's past criminal record. Overall, jurors tend to use that information to increase the probability of a guilty verdict.
  5. Jurors' use of inadmissible evidence (esp., about criminal record) depends on:
    a. Similarity of previous offenses to the one currently being tried: I similar, jurors more likely to use it as indication of guilt (**Sealy & Cornish, 1973; Wissler & Sakes, 1985**)
    b. Strength of existing case evidence: Jurors use inadmissible evidence if other case evidence weak, but not when it is strong
    c. In mock jury cases, inadmissible evidence more likely to be used in less serious crimes (vandalism) than in more serious ones (rape, murder.) See **Rind et al (1995); Sealy & Cornish (1973)**
    d. The reason for admissibility: E.g., if confession tape not admissible because it is prejudicial, it will still be used; if inadmissible because it is not fully audible, jurors may ignore it in their decision making. (**Kassin & Sommers, 1997**)

IV. Research on the comprehension of judicial instructions
   A. Findings from research with actual jurors
      1. Questioning jurors
         a. **Reifman, Gusick & Ellsworth (1992)**: Mailed out questionnaires to Michigan jurors after their service. Received responses from 40% (224, including 140 who had served on a jury). Asked T/F questions about procedural duties and substantive law related to serious crimes. Found:
            (1) did better if instructed than if not
            (2) did better if served on criminal rather than civil jury
            (3) best performance was still below chance (47.8%)
            (4) jurors on criminal cases no better on criminal issues (41%) than on issues unrelated to their cases
            (5) two jurors who were lawyers only scored 70% correct.
            (6) jurors on juries that asked from help from judges did better, but no effect of addresses by counsel
         b. **Jackson (1992)**: Surveyed all jurors in England and Wales over two-week period in 1992. Found that:
            (1) 65% reported understanding 'all' of judges instructions; 25% 'most', only 10% 'some' or 'none'.
            (2) 97% say understood summing up, and 93% said is was useful.
            (3) but of the 32% of jurors told to disregard evidence, 84% said they understood why, but only 68% said they had been successful in doing so.
         c. **Young, Cameron & Tinsley (2001)**: Questioned jurors in 48 trials in New Zealand. Found that
            (1) 85% reported that the judge's instructions were 'Clear'
            (2) 80% found them 'helpful'
            (3) but some jurors in 72% of trials misunderstood some aspect of the relevant law, most often the legal requirement for a finding of guilt on one or more of the charges.
            (4) New Zealand now undertaking reforms to jury system (**Eames, 2003**
         d. **Rose, Chopra & Ogloff (2001)**: Did semi-structured interviews with called jurors in B.C. over previous two years. Found that juror comprehension of judicial instructions low, and in line with lab studies.
      2. Field studies
         a. **Saxton (1998)**: Real jurors (and counsel and judges) in Wyoming complete questionnaire before being discharged. Found that 97% of jurors reported understanding instructions, which they considered helpful (94%). But scores on comprehension questions score only 74% for criminal jurors, and 58% for civil jurors.
         b. Juries' questions: **Severance & Loftus (1982, 1984)** studied jurors questions to judges in Washington over period of years. Found jurors had most trouble with concepts of 'intent' and 'reasonable doubt', and that judges generally just referred them back to the original instructions.

   B. Comprehension studies in simulated trials

1. **Jones & Myers (1979)**: Using instructions used in an actual Canadian murder/manslaughter case, found 62% comprehension on 11 questions about instructions, and 72% comprehension after redrafting the instructions based on psycholinguistic principles.
2. **Rose & Ogloff (2001)**: Using Canadian instructions on conspiracy and the co-conspirator exception to the hearsay rule found comprehension in the 60-70% range, and equal confidence in correct and incorrect answers. (Suggesting that allowing jurors to ask questions when they are unsure may not help, since they are equally sure of things they have wrong.)
3. **Diamond (1993)**: 40-68% of Participants got application questions correct.
4. **Ogloff (1998)**: Day-long study of 500 jury-eligible individuals shown 2.5 hour video trial based on actual case, including jury instructions. Jurors spent only 8% of jury time discussing instructions - about 75 seconds in a maximum two-hour deliberation time. When they tried to define the judicial instructions (which they did rarely), they were correct only about **60%** of the time.

V. Research on strategies/Mechanisms to Increase Comprehension of Judicial Instructions

A. "Plain language" or redrafted instructions: Several studies find that it raises comprehension of instructions, but not hugely:
   1. **Frank & Applegate (1998)**: Penalty-phase capital case instructions: 68% vs 50%
   2. **Elwork et al (1977)**: Michigan negligence instructions: 60% vs 81%
   3. **Jones & Myers (1979)**: Canadian murder/manslaughter case: 62% vs 72%

B. Juror note-taking: In Canada, trial judge can decide whether jurors allowed to take notes or not.
   1. Concerns about note-taking:
      a. Note-taking is an acquired skill, so note-takers will appear more knowledgeable and have more influence in deliberations
      b. Notes may not be accurate, focusing on less important aspects of evidence
      c. Note-takers may distract other jurors
      d. Notes may be incomplete; perhaps more at the beginning and less at the end of a trial
   2. Jurors want to take notes (**Cutler & Hughes, 2001**): ~40% of jurors surveyed in North Carolina were allowed to take notes, and ~39% of the remainder thought it would have been helpful.
   3. **Heuer and Penrod (1988, 1994)**: Field experiments in Wisconsin and across U.S. Found that jurors took about .6 pages of notes per hour in criminal and civil trials, and notes were accurate, did not favor one side or the other
   4. **Penrod & Heuer (1997)** reviewed note-taking research and concludes that:
      a. Jurors can keep up with the evidence as they take notes
      b. Jurors who take notes do not distract those who don't
      c. Jurors' notes are an accurate record of the trial.
      d. Juror's notes serve as memory aid
      e. Jurors' notes do not distort their view of the case
      f. Jurors do not overemphasize noted evidence over evidence they did not note
      g. Jurors who took notes do not have an undue influence over those who did not.

C. Asking questions: May be allowed by the judge. Right now, jurors may submit questions for a witness to the judge, who will ask them after lawyers have finished questioning the witness. **Penrod & Heuer (1997)** conclude re asking questions that:
1. Jurors do not ask legally inappropriate questions
2. Jurors do not become advocates
3. Jurors do not draw inappropriate inferences about unanswered questions.
4. Jurors questions promotes juror understanding of facts and issues.
5. Juror questions do not clearly help get to the truth.
6. Juror questions do not increase participants' satisfaction with the trial or verdict.

D. Preinstruction of juries
1. **Cruse & Browne (1987)**: In simulated larceny trial, timing of instructions made no difference, but frequency did: More instructions better.
2. **Heuer & Penrod (1989)**: In field study, jurors expected pre-instructions to be helpful, and found them even more helpful than they had expected, though no apparent influence on actual recall of instructions or understanding of trial procedure.
3. **Kassin & Wrightsman (1979)**: Found that preinstruction on requirements of proof, presumption if innocence, and reasonable doubt reduced conviction rates and ratings of probability of factual guilt. Preinstructions appeared to cause adoption of presumption of innocence.

E. Providing juries with copies of the jury charge: Jurors think it would be helpful, say it was helpful if they received them, but no evidence that it improves understanding or recall of instructions.

F. **Ogloff (1998)**: (Canadian) Day-long study of 500 jury-eligible individuals shown 2.5 hour video trial based on actual case, including jury instructions. Tried several alternative strategies for increasing comprehension of jury instructions:
1. "plan language" instructions
2. juror note-taking
3. pre- and post-trial instructions
4. proving copies of judges instructions
5. a decision-tree jury deliberation model
6. none were especially effective, but pre- as well as post-trial instructions and the decision-tree model did increase comprehension slightly.

VI. Judge - Jury Agreement:

A. **Kalven & Zeisel (1966)**: Classic work with Chicago jury project. Found that judges agreed with jurors' verdicts in 75.4% of cases. With most disagreements (16.9%) occurring when judge would have convicted, but jury did not. In 1,083 cases in which jury acquitted, judge would have convicted in 57% (604) of those cases.

B. **Saxton (1998)** Wyoming study: Criminal judges agreed with their jurors' verdict 76% of the time, but only 50% of the time in civil cases. Virtually all judges (100%) in both types of cases believed that their jurors understood the relevant law.

C. **Young, Cameron & Tinsley (2001)**: New Zealand study. Met with judge at close of trial (not a recall study as with others above), and found that judges disagreed with jurors' verdict 50% of the time - chance.

References

Bull, R. & Rumsey, N. *The Social Psychology of Facial Appearance*. New York: Springer-Verrlag, 1988.

Costanzo, M. *Psychology Applied to Law*. Belmont, CA: Wadsworth, 2003.

Cruse, D. & Browne, B.A. Reasoning in a jury trial: The influence of instructions. *Journal of General Psychology*, 1987, 114, 129-133.

Cutler, B.L. & Hughes, D.M. Judging jury service: Results of the North Carolina administrative office of the course juror survey. *Behavioral Sciences and the Law*, 2001, 19, 305-320.

Cutler, B.L., Penrod, S.D. & Stuve, T.E. Juror decision making in eyewitness identification cases. *Law and Human Behavior*, 1988, 12, 41-55.

DeLuca, A. Tipping the scales of justice: The effects of pretrial publicity. Unpublished Masters' thesis, Iowa State University, Ames Iowa.. (1979):

Devine,. D.J., Clayton, L.D., Dunford, B.B., Saying, R. & Pryce, J. Jury decision making: 45 years of empirical research on deliberating groups. *Psychology, Public Policy, & Law*. 2001, 7, 622-727.

Diamond, S.S. Instructing on death: Psychologists, juries and judges. *American Psychologist*, 1993, 48, 423-434.

Eames, J. Towards a better direction: Better communication with jurors. *Australian Bar Review*, 2003, 24, 35-78.

Elwork, A., Sales, B.D. & Alfini, J.J. Juridic decisions: In ignorance of the law or in light of it? *Law and Human Behavior*, 1977, 1, 163-189.

Elliot, R., Farrington, B. Manheimer, H. Eyewitness credible and discredible. *Journal of Applied Social Psychology*, 1988, 44, 20-33.

Ellsworth, P.E. & Mauro, R. Psychology and Law. In D.T. Gilbert & S.T. Fiske (Eds.) *The Handbook of Social Psychology* (4th Edition), vol. 2, pp 684-732). New York: McGraw-Hill, 1998.

Frank, J. & Applegate, B.K. Assessing juror understanding of capital-sentencing instructions. *Crime and Delinquency*, 1998, 44, 412-433.

Hastie, R. Algebraic models of juror decision processes. In R. Hastie (Ed.) *Inside the Juror: The Psychology of Juror Decision-Making*. New York: Cambridge University Press, 1993.

Hastie, R., Penrod, S.D. & Pennington, N. *Inside the Jury*. Cambridge, MA: Harvard University Press, 1983.

Heuer, L. & Penrod, S.D. Increasing jurors' participation in trials: A field experiment with jury notetaking and question asking. *Law and Human Behavior*, 1988, 12, 231-261.

Heuer, L. & Penrod, S.D. Instructing jurors: A field experiment with preliminary and written instructions. *Law and Human Behavior*, 1989, 13, 409-430.

Heuer, L. & Penrod, S.D. Trial complexity: A field investigation of its meaning and its effects. *Law and Human Behavior*, 1994, 18, 29-51.

Huntley, J.E. & Costanzo, M. Sexual harassment stories: Testing a story-mediated model of juror decision-making in civil litigation. *Law and Human Behavior*, 2003. 27, 29-51.

Hvistendahl, J.K. The effect of placement of biasing information. *Journalism Quarterly*, 1979, 56, 863-865.

Imrich, D.J., Mullin, C. & Linz, D. Measuring the extent of pretrial publicity in major American newspapers: A content analysis. *Journal of Communication*, 1995, 45(3), 94-117.

Jackson, J. Juror decision-making and the trial process. In G. Davis & S. Lloyd-Bostock (Eds.) *Psychology, law, and criminal justice: International developments in research and practice*. 1992), Oxford, UK: DeGuyter.

Jones, C.S. & Myers, E. Comprehension of jury instructions in a simulated Canadian court. In Law Reform Commission of Canada (Eds.) *Studies on the jury*. Ottawa: Law Reform Commission of Canada, 1979.

Kalven, H. & Zeisel, H. *The American jury*. Boston: Little, Brown, 1966.

Kassin, S.M. & Neumann, K. On the power of confession evidence: An experimental test of the fundamental difference hypothesis. *Law and Human Behavior*, 1997, 21, 469-484.

Kassin, S.M. & Sommers, S.R. Inadmissible testimony, instructions to disregard, and the jury: Substantive vs procedural considerations. *Personality and social Psychology Bulletin*, 1997, 23, 1046-1054.

Kassin, S.M. & Sukel, H. Coerced confessions and the jury: An experimental test of the 'harmless error' rule. *Law and Human Behavior*, 1997, 21, 27-46.

Kassin, S.M. & Wrightsman, L.S. On the requirements of proof: The timing of judicial instructions, and mock juror verdicts. *Journal of Personality and Social Psychology*, 1979, 37, 1877-1887.

Kassin, S.M. & Wrightsman, L.S. The construction and validation of a juror bias scale. *Journal of Research in Personality*, 1983, 17, 423-442.

Kerr, N.L., MacCoun, R..J. & Kramer, G.O. Bias in judgment: Comparing individuals and groups. *Psychological Review*, 1996, 103, 687-719.

Kerwin, J. & Shaffer, D.R. Mock jurors versus juries: The role of deliberations in reactions to inadmissible testimony. *Personality and Social Psychology Bulletin*, 1994, 20, 153-162.

Kovera, M.B., McAuliff, B.D. & Hebert, K.S. Reasoning about scientific evidence: Effects of juror gender and evidence quality on juror decisions in a hostile work environment case. *Journal of Applied Psychology*, 1999, 84, 362-375.

Kramer, G.P., Kerr, N.L. & Carroll, J.S. Pretrial Publicity, judicial remedies and jury bias. *Law and Human Behavior*, Vol. 14, No. 5, 1990:

Landsman, S. & Rakos, R.F. A preliminary inquiry into the effect of potentially biasing information on judges and jurors in civil litigation. *Behavioral Sciences and the Law*, 1994, 12, 113-126.

Loo, Robert. A critical examination of rhe Rubin And Peplau Belief In A Just World Scale (pp. 125-145). *Advances in Psychology Research*, Vol. 31, pp. 125-145.

MacCoun, R.J. & Kerr, N.L. Asymmetric influence in mock jury deliberations: Jurors' bias for leniency. *Journal of Personality and Social Psychology*, 1988, 54, 21-33.

Maeder, E.M., Yamamoto, S. & Saliba, P. (2015). The influence of defendant race and victim physical attractiveness on juror decision-making in a sexual assault trial. *Psychology, Crime & Law*, Vol. 21(1), 62–79.

Mazzella, R. & Feingold, A. The effects of physical attractiveness, race, socioeconomic status, and gender of defendants and victims on judgments of mock jurors: A meta-analysis. *Journal of Applied Social Psychology*, 1994, 24, 1315-1344.

Miene, P., Park, R. & Borgida, E. Juror decision making and the evaluation of hearsay evidence. *Minnesota Law Review*, 1992, 76, 51-94.

Narby, D.J., Cutler, B.L. & Moran, G. A meta-analysis of the association between authoritarianism and jurors' perceptions of defendant culpability. *Journal of Applied Psychology*, 1993, 78. 34-42.

Nuñez, N., Schweitzer, K., Chai, C.a. & Myers, B. (2015) Negative Emotions Felt During Trial: the Effect of Fear, Anger, and Sadness on Juror Decision Making. Applied Cognitive Psychology, 29, 200–209.

Ogloff, J.R.P. *Judicial instructions and the jury: A comparison of alternative strategies. Final report.* 1998. Vancouver, B.C. British Columbia Law Foundation.

Ogloff, J.R.P. & Vidmar, N. The impact of pretrial publicity on jurors: A study to compare the relative effects of television and print media in a child sex abuse case. *Law & Human Behavior*, 1994, 18(5), 507-525.

Olsen-Fulero, L. & Fulero, S.M. Commonsense rape judgments: An empathy-complexity theory of rape juror story making. *Psychology, Public Policy, and Law*, 1997, 3, 402-427.

Otto, A.L., Penrod, S. & Dexter, H. The biasing impact of pretrial publicity on juror judgments. *Law and Human Behavior*, 1994, 18, 453-470.

Pennington, N. & Hastie, R. Juror decision-making models: The generalization gap. *Psychological Bulletin*, 1981, 89, 246-287.

Pennington, N. & Hastie, R. Evidence evaluation in complex decision making. *Journal of Personality and Social Psychology*, 1986, 51, 242-258.

Pennington, N. & Hastie, R. Explaining the evidence: Tests of the story model for juror decision making. *Journal of Personality and Social Psychology*, 1992, 62, 189-206.

Penrod, S. & Otto, A.L. *Pretrial publicity and juror decision making: Assessing the magnitude and source of prejudicial effects.* Paper presented at the Third European Conference on Law and Psychology, Oxford, UK, 1992.

Rakos, R.F. & Landsman, S. Researching the hearsay rule: Emerging findings, general issues, and future directions. 1992, *Minnesota Law Review*, 76, 655-681.

Reifman, A., Gusick S.M. & Ellsworth, P.C. Real jurors' understanding of the law in real cases. *Law and Human Behavior*, 1992, 16, 539-554.

Rind, B., Jaeger, M. & Strohmetz, D.B. Effect of crime seriousness on simulated jurors' use of inadmissible evidence. *Journal of Social Psychology*, 1995, 135, 417-424.

Rose, V.G. & Ogloff, J.R.P. Evaluating the comprehensibility of jury instructions: A method and an example. *Law and Human Behavior*, 2001, 25, 409-431.

Rose, V.G., Chopra, S.R. & Ogloff, J.R.P. *The perceptions and reactions of real Canadian criminal jurors.* Paper presented at annual convention of CPA, Ste-Foy, Quebec, 2001.

Ruva, C.L. & Guenther, C.C. (2015). From the Shadows Into the Light: How Pretrial Publicity and Deliberation Affect Mock Jurors' Decisions, Impressions, and Memory. Law and Human Behavior, Vol. 39 (3), 294–310.

Saks, M.J. & Wissler, R.L. Legal and psychological bases of expert testimony: Surveys of the law and of jurors. *Behavioral Sciences and the Law*, 1984, 2, 435-449.

Saxton, B. How well do jurors understand jury instructions? A field test using real juries and real trials in Wyoming. *Land and Water Law Review*, 1998, 33, 59-189.

Sealy, A.P. & Cornish, W.R. Juries and the rules of evidence. *Criminal Law Review*, 1973, 208-223.

Severance, L.J. & Loftus, E.F. Improving the ability of jurors to comprehend and apply criminal jury instructions. *Law and Society Review*, 1982, 17, 153-197.

Severance, L.J. & Loftus, E.F. Improving criminal justice: Making jury instructions understandable for American jurors. *International Review of Applied Psychology*, 1984, 33, 97-119.

Stasser, G. Information salience and the discovery of hidden profiles by decision-making groups: A 'thought experiment'. *Organizational Behavior and Human Decision Processes*, 1992, 52, 156-181.

Steblay, N.M., Besirevic, J., Fulero, S.M. & Jimenez-Lorente, B. The effects of pre-trial publicity on juror verdicts: A meta-analytic review. *Law and Human Behavior*, 1999, 23, 219-235.

Sue, S., Smith, R.E. & Caldwell, C. Effects of inadmissible evidence on the decisions of simulated jurors: A moral dilemma. *Journal of Applied Social Psychology*, 1973, 3, 345-353.

Sommers, S.R. & Ellsworth, P.C. Race in the courtroom: Perceptions of guilt and dispositional attributions. *Personality and Social Psychology Bulletin*, 2000, 26, 1367-1379.

Wetmore, S.A., Neuschatz, J.S. & Gronlund, S.D. (2014). On the power of secondary confession evidence. *Psychology, Crime & Law*, Vol. 20, No. 4, 339 357.

Wissler, R.L. & Saks, M.J. On the inefficacy of limiting instructions: When jurors use prior conviction evidence to decide on guilt. *Law and Human Behavior*, 1985, 9, 37-48.

Young, W., Cameron, N. & Tinsley, Y. *Juries in criminal trials* (Report 69), Wellington, NZ: New Zealand Law Commission, 2001.

Psychological Assessment: Competence & Criminal Responsibility

I. Introduction

   A. Now leaving the realm of empirical psychology for the realm of clinical psychology, where data are much scantier. Except for:
      1. Statistics on case types
      2. Information on test reliability, validity

   B. Assessment involves looking at the present, into the past, and into the future
      1. Insanity and criminal responsibility - past
      2. Fitness for trial - present
      3. Risk of violence - future

   C. We will look at these assessments in the order in which they might occur during an actual legal case: Fitness, Criminal Responsibility, Future Risk of Violence

II. Competence to Stand trial - Adjudicative Competence

   A. According to the Canadian Criminal code (Bill C-30, 1992) a defendant is unfit to stand trial by virtue of a mental disorder if he:
      1. "*Is unable on account of mental disorder to conduct a defence at any stage of the proceeding before a verdict is rendered or to instruct counsel to do so, and in particular, unable on account of mental disorder to a) understand the nature or object of the proceedings b) understand the possible consequences of the proceedings, or c) communicate with counsel.*"

   B. In U.S., use the criterion laid out by Supreme Court in *Dusky v United States* (1960):
      1. "*The test must be whether he has sufficient present ability to consult with his attorney with a reasonable degree of rational understanding and a rational as well as factual understanding of proceedings against him*"

   C. Domains and sub-domains of adjudicative competence
      1. **Capacity to comprehend and appreciate the charges or allegations**
         a. Factual knowledge of the charges (ability to report charge label)
         b. Understanding the behaviors to which the charge refers
         c. Comprehension of police version of events
      2. **Capacity to disclose to counsel pertinent facts, events, and states of mind**
         a. Able to provide reasonable account of behavior around time of offense
         b. Able to provide information about state of mind around time of offense

- c. Able to provide account of behavior of others around time of offense
- d. Able to provide account of police behavior
3. **Capacity to comprehend and appreciate range and nature of possible penalties**
   - a. Knowledge of penalties that could be imposed (knowledge of sentence associate with charge, e.g., '5 to life')
   - b. Comprehension of seriousness of charges and possible sentences
4. **Basic knowledge of legal strategies and options**
   - a. Understanding of the meaning of alternative please (e.g., guilty and NCRMD)
   - b. Knowledge of the plea bargaining process
5. **Able to engage in reasoned choice of legal strategies and options**
   - a. Able to comprehend legal advice
   - b. Able to participate in planning a defense strategy
   - c. Able to appraise likely outcome (i.e., likely disposition for own case)
   - d. Understanding of implications of guilty plea of plea bargain (e.g., rights waived after guilty plea)
   - e. Able to make reasoned choice of defense options without distortion attributable to mental illness
6. **Capacity to understand the adversary nature of the proceedings**
   - a. Understanding role of courtroom personnel (i.e., judge, jury, crown attorney)
   - b. Understanding courtroom procedure (i.e., basic sequence of trial events)
7. **Capacity to show appropriate courtroom behavior**
   - a. Understanding of appropriate courtroom behavior
   - b. Ability to manage own emotions and behavior in the courtroom
8. **Capacity to participate in trial**
   - a. Able to track events as they occur
   - b. Able to challenge witnesses (i.e., to recognize distortions in witness testimony)
9. **Capacity to give relevant testimony**
10. **Appropriate relationship with counsel**
    - a. Awareness that counsel is an ally
    - b. Appreciation of attorney-client privilege; trust and confidence in one's attorney
    - c. Confidence in attorneys in general
11. **Medication issues**
    - a. Able to track proceedings and communicate with counsel given level of sedation
    - b. Possible negative effects of medication on courtroom behavior
    - c. Bases for treatment refusal

D. Who Can Assess Fitness?
  1. In Canada, only medical practitioners can conduct fitness assessments - even if they have no background in psychology or psychiatry
  2. In many other countries, including most states in the U.S., psychologists can also conduct such evaluations.
  3. Normally 5 days is limit for psychological evaluation, but can be extended to 30 days, with detention not to exceed 60 days.

E. Basically four kinds of tests used in fitness assessments:
  1. General tests for psychopathology (MMPI, MCMI)
  2. Neuropsychological batteries to detect brain damage (Luria-Nebraska, Halstead-Reitan)

3. Intelligence tests to detect retardation (WAIS, Stanford-Binet)
4. Tests specific for the fitness criteria (MacCAT-CA, FIT-R)

F. Fitness Instruments generally: Although all assessments involve a personal interview with client, many if not all assessments include use of one or more standardized tests.
1. Personal disclosure: I am not trained in, and have not administered or interpreted any of the tests we will mention.
2. Given my credentials, I cannot even purchase copies of many of them, so must describe them purely from whatever public literature exists. Often, that does not describe the specific items on the test.
3. Only reliability/validity information is widely available.

G. Lally (2003). Surveyed 64 diplomates in forensic psychology. Found the following with respect to test usage in CST assessments:
1. **Acceptable**:
   a. MacCAT-CA - 90%
   b. WAIS III - 90%
   c. CAI - 85%
   d. CST - 77%
   e. MMPI-2 - 73%
   f. GCCT - 65%
   g. Halstead-Reitan Neuropsychological Battery- 64%. Eight tests that evaluate range of nervous system and brain functions, including: visual, auditory, and tactual input; verbal communication; spatial and sequential perception; the ability to analyze information, form mental concepts, and make judgments; motor output; and attention, concentration, and memory. Typically used to evaluate individuals with suspected brain damage.
   h. IFI-R - 62%
   i. Stanford-Binet-Revised - 54%
   j. ~~PAI (**Personality Assessment Inventory**)- 52%~~
   k. Luria-Nebraska Neuropsychological Battery - 50%. Measures neuropsychological functioning in several areas: Motor skills, language abilities, intellectual abilities, nonverbal auditory skills, and visual-spatial skills. Used as screening tool for significant brain injury, or to learn more about known injuries.
2. **Unacceptable**
   a. Projective drawings - 87%
   b. TAT - 77%
   c. Sentence completion tests - 69%
   d. Rorschach - 60%
   e. 16PF - 58%
   f. MCMI-II - 50%

H. Lots of different instruments used:
1. **Georgia Court Competency Test (GCCT)**: Developed by Dr. Robert W. Wildman (Ph.D.) of the Central State Hospital in Milledge, GA (**Wildman et al, 1978**); **Wildman et al (1990)**
2. **Georgia Court Competency Test - Mississippi Version Revised (GCCT-MSH)**:

Ustad et al (1996)
3. **Competency Screening Test (CST):** Lipsitt, Lelos & McGarry (1971); McGarry (1973) See Pozzulo, p. 266
4. **Competency Assessment Instrument (CAI):** Grisso (1986, 1988)
5. **Interdisciplinary Fitness Interview (IFI):** Golding & Roesch (1981); Golding et al 1984) - See Melton et al, p 756-57. Pozzulo, p. 267
6. **Computer-Assisted Determination of Competency to Proceed** (CADCOMP): **Bernard et al (1991)**
7. **MacArthur Competence Assessment Tool - Criminal Adjudication** (MacCAT-CA):
   a. Developed with decade of funding from MacArthur Foundation and NIMH. See Melton et al, p 149. See Pozzulo, p. 267; tested in a number of studies in the 1990s
   b. 22-item measure administered as semistructured interview, taking 25-45 minutes
   c. Yields quantitative indices of:
      (1) Understanding
      (2) Appreciation
      (3) Reasoning
   d. To evaluate Understanding and Reasoning, person reads vignette about charge of aggravated assault against Fred as a result of bar fight with Reggie. Then asked questions about Fred's situation and options.
   e. To test Appreciation, defendant asked questions about what might happen to him as a result of the proceedings.
8. **Evaluation of Competency to Stand Trial - Revised** (ECST-R): See Melton. P. 152-53.
9. **Fitness Interview Test** (Revised) - FIT-R (**Roesch, Zapf, Eaves & Webster, 1998**). Specifically Canadian, so not mentioned by U.S. practitioners.
   a. A semistructured clinical interview of 16 items that takes about 30 minutes to administer. Originally developed in 1984.
   b. As revised, is divided into 3 sections, paralleling three legal criteria for fitness to stand trial outlined in the 1992 amendments of the Criminal Code of Canada:
      (1) Understanding the nature and object of the proceedings
      (2) Understanding the possible consequences of the proceedings
      (3) The ability to communicate to counsel
   c. FIT items rated on a 3-point scale. Though performance on individual items in each section is considered in determining section rating, decisions not made on basis of cut-off score. Instead, section ratings constitute separate judgment based on the severity of impairment and its perceived importance.
   d. **Zapf and Roesch (1997)** compared decisions made by FIT with decisions made in institution-based evaluation of fitness. Found:
      (1) FIT correctly identified 49 of 57 male defendants in their sample (86%) as either fit or unfit.
      (2) Emphasized that screening instruments must not make false-negative errors (that is, call a defendant "fit" who is truly unfit), and FIT made no false-negative errors.
      (3) FIT has excellent agreement with institution-based decisions (**Zapf et al, 2001**), and appropriate levels of agreement with other fitness assessment instruments (**Zapf & Roesch, 2001**).

I. How is Fitness Restored?
    1. Medication usually
    2. But can defendant refuse medication? Courts may intervene to require medication if the judgment is that the individual is not competent to make that decision on his own.

J. What Happens After a Finding of Unfitness?

III. Mental State at the Time of the Offence

   A. Introduction: Presumptions in Canada's legal system
      1. *Actus reus* and *mens rea*
         a. "The terms actus reus and mens rea are derived from the principle stated by Edward Coke, namely, *actus non facit reum nisi mens sit rea*, which means: "an act does not make a person guilty unless (their) mind is also guilty", i.e., the general test is one that requires proof of fault, culpability or blameworthiness both in behaviour and mind.
         b. Once the *actus reus* has been established in a conventional offence, there must be a concurrence of both *actus reus* and *mens rea*.
      2. Note that 'insane' and 'insanity' are legal not psychological or psychiatric terms. Neither is in DSM, and do not refer to recognized categories of mental illness.
      3. They refer to the legal concept of *mens rea*: That you must have intent to commit illegal act to be responsible. In insanity, *mens rea* is missing because the individual is unable to appreciate quality of actions, or unable to stop actions from taking place.

   B. History of insanity
      1. References to criminal insanity in Roman law: Since mental illness ("madness") a punishment itself, we should be lenient with the insane.
      2. In the 13th, prominent writers argued that since the mentally ill lacked 'the will to harm', they were excused of criminal responsibility for their acts.
      3. In 1603 Sir Edward Coke argued that since the mentally ill person 'did not know what he did', he could not have any criminal intent (*mens rea*)
      4. The 'wild beast test' of 1724 stated that persons acting under the influence of 'animal reflexes' rather than moral choice were not responsible for criminal actions.
      5. Eventually additional criteria came into play, most notably the ability to distinguish right from wrong.

   C. Canadian ideas of insanity (and mostly those in U.S. as well) come from two pivotal British cases:
      1. **Hadfield case (1800)**: James Hadfield, who had suffered a brain injury in 1794 at the battle of Tourcoing. Suffered multiple sabre wounds to the head. the army, tried to assassinate King George III in the belief that he would be hanged, and thereby hasten the second coming of Jesus Christ.
         a. Insanity standard at the time was that defendant must be "lost to all sense …

incapable of forming a judgement upon the consequences of the act which he is about to do". Because Hadfield had planned the assassination, he did not meet the standard.

b. Thomas Erskine argued in treason trial that Hadfield out of touch with reality, and his delusions, though "unaccompanied by frenzy or raving madness" constituted insanity.

c. Judge ordered verdict of acquittal, but ordered that prisoner not be discharged - as was typically the outcome of successful insanity pleas.

d. Public outrage at the verdict led Parliament to pass Criminal Lunatics Act that provided for detention of insane defendants. Hadfield was detained in Bethlehem Royal Hospital ('Bedlam') for the rest of his life.

e. **R. v. M'Naughten (1843)**: This is the more famous case.

f. M'Naughten had delusion of a conspiracy involving (among others) the Pope and the Tory government. And decided to murder the Prime Minister, Robert Peel. (For whose first name the term 'Bobby' for British policeman came after he founded Metropolitan Police as Home Secretary in 1829) In January, 1843, he shot and killed Edward Drummond, Peel's Secretary.

g. Found not guilty by reason of insanity, and the case established the primary considerations in such a plea:

  (1) A person is deemed insane if: "*at the time of committing the act, he was laboring under such a defect of reason from disease of the mind as not to know the nature and quality of the act he was doing, or if he did know it, that he did not know what he was doing was wrong.*" This test is also commonly referred to as the "right/wrong" test.

h. Three key aspects of the M'Naughten definition:

  (1) Suffering from a defect of reason due to disease of the mind.

  (2) Ignorant of the nature and quality of the act; or

  (3) Unaware, or unable to determine, that the act was wrong.

i. In the U.S, about half the states (22 to 26, depending on who's counting) still the standard as modified by the American Law Institute (1962): Insane "if at the time of his conduct as a result of mental disease or defect he lacks substantial capacity either to appreciate the criminality (wrongfulness) of his conduct **or to conform his conduct to the requirements of law**."

j. This rule (as in bold section above) is less restrictive than M'Naughten rule. Some states have modified M'Naughten rule to include defendant suffering from irresistible impulse which prevents him from being able to stop himself from committing an act that he knows is wrong.

k. Interestingly, in three Western states (Montana, Idaho, Utah) insanity is not a recognized legal defense.

IV. Assessing Insanity: What Happens to a Defendant Found NCRMD?

    A. Raising the issue of insanity
        1. Defendant can raise it as a defense, and Crown can argue it
        2. Crown can raise following a guilty verdict (if Crown thinks defendant requires psychiatric treatment in a mental facility)
        3. Standard of proof is "*beyond a balance of the probabilities*" (rather than beyond a reasonable doubt)

    B. Psychiatric assessment required. A number of different assessment tools used:
        1. Structured clinical interviews:
            a. **SADS** (Schedule for Affective Disorders and Schizophrenia): (Spitzer & Endicott, 1978).
        2. Projective personality tests:
            a. **Rorschach Inkblot Test**: According to **Borum and Grisso (1995)** as cited by **Rogers & Sewell (1999)**, the Rorschach is the most commonly used projective test in insanity cases. Borum & Grisso say that one third of forensic psychologists and psychiatrists use the Rorschach as part of insanity evaluations.
            b. TAT (**Thematic Apperception Test**)
        3. Objective personality inventories:
            a. MMPI/MMPI-2 (**Minnesota Multiphasic Personality Inventory**)
            b. MCMI (**Millon Clinical Multiaxial Inventory**)
        4. Forensic instruments:
            a. Richard **Rogers Criminal Responsibility Assessment Scales** (R-CRAS; **Rogers, 1984**). Has five scales, each containing 30 items. Each item scored from 0 to 6, with high values indicating greater severity:
                (1) Patient reliability
                (2) Organicity
                (3) Psychopathology
                (4) ~~Cognitive control~~
                (5) Behavioral control
        5. Cognitive and intelligence tests:
            a. WAIS-R
            b. WMS-R (**Wechsler Memory Scale-Revised**)
        6. Neuropsychological tests:
            a. Luria-Nebraska Neuropsychological Battery
            b. Halstead-Reitan Neuropsychological Battery
        7. **Holub (1992)** mentions that 67% of clinical psychologists mention two other tests they use as part of forensic assessment:
            a. Wechsler Adult Intelligence Scale - Revised (WAIS-R)
            b. Bender Visual-Motor Gestalt Test
        8. Etc.
        9. Psychologists vs psychiatrists:

Test	% Psychologists Using	% Psychiatrists Using
MMPI/MMPI-2	94%	80%
MCMI	32%	17%
Rorschach	32%	30%
TAT	8%	10%
WAIS	78%	57%
WMS-R	16%	0%
Bender VMGT	12%	20%
R-CRAS	41%	10%
SIRS	12%	0%

V. Interview techniques and tools

   A. Clinical interview
      1. Will certainly begin with simple medical/psychiatric history: Any previous problems with mental illness. When? What? What done?
      2. Want to minimize clinician influence on defendant's recall
      3. Best done with simple, open-ended questions:
         a. "What was happening?"
         b. "What were you aware of?"
         c. "What caught your attention?"
         d. "What thoughts do you remember?"
         e. "What were you feeling?"
      4. The day of the offense: Start with waking up, and move through the day capturing as much as possible of the defendant's actions, thoughts, and feelings. If structure needed to the narrative, ask more specific questions:
         a. "What time did you wake up?"
         b. "What happened next?"
         c. "What were you thinking about as you had breakfast?"
      5. Days preceding the offense: Collect salient thoughts, any changes in the defendant or his environment. Need to address the questions "Why then?" with respect to criminal actions.
      6. Rogers suggests several strategies for getting more recall, and disrupting memorized or (seemingly) rehearsed narratives:
         a. Ask subject to re-experience the day or time, as though a video camera were recording it. What would the camera have recorded?
         b. Ask about events in reverse chronological order: What happened just before that? (Rogers notes that using this technique he was able to recover more than a day of lost memories from a self-reported amnesiac.)

7. Structured clinical interviews: **SADS** (Schedule for Affective Disorders and Schizophrenia):
   a. (Spitzer & Endicott, 1978). Goes beyond DSM-IV to assess clinical characteristics and associate features of psychotic and mood disorders. Also partly covers anxiety abuse and substance disorders.
   b. Initially designed for assessments of present rather than past mental states
   c. Part I addresses current episode from two perspectives: the worst time and the past week or so
   d. Part II designed to assess past episodes, nature of diagnoses and treatments
   e. Takes 90 to 150 minutes to administer
8. Rogers uses **SADS**, but modifies it so that Part I addresses not worst period and current time, but the time of the offense, and the current time. Likes SADS over **SCID** (Structured Clinical Interview of DSM-IV Disorders; First et al, 1997) because:
   a. SADS scores symptoms severity; SCID focuses on symptom presence or absence
   b. High reliability for key psychotic and mood symptoms
   c. Existence of detection strategies for feigned disorders (from Rogers, 1997)
9. *"The centerpiece of insanity evaluations is the systematic use of interview methods for the establishment of retrospective diagnoses, prominent symptoms, and psychological impairment."* (Rogers, p. 119; in Jackson)

VI. Introduction to Projective tests

   A. Are based on, and stem directly from the assumptions about personality of the psychodynamic approach.

   B. Assume that the most important aspects of an individual's personality are determined by unconscious factors, which the tests are designed to tap.

   C. Tests assume that when the individual is presented with an ambiguous situation or stimulus, or with a task for which there is no clear correct answer, the individual will <u>project</u> material from the unconscious into his or her response, so that the answer will reflect the subject's inner feelings, desires, etc.

   D. Projective tests use questions or tasks that involve minimal structure, and minimal restriction of a subject's responses.

   E. Instructions for administration generally fairly standardized, but scoring is often less so, and interpreting the subject's scores or responses is as much art as science, and different experts may disagree on the interpretation of a given set of responses. Is thought to require considerable training, insight and clinical experience.

VII. **The Rorschach Inkblot Test**

　　A. History of the test
　　　　1. Developed by Swiss psychiatrist Hermann Rorschach in 1921.
　　　　2. Intended it for use in the clinical diagnosis of psychiatric patients.
　　　　3. In 1985 or so, rated as the fourth most popular in use by psychodiagnosticians **(Lubin et al, 1984)**

　　B. Description of the test.
　　　　1. Consists of a series of 10 symmetrical inkblots, reproduced from those used by Rorschach (available from Hans Huber in Bern.)
　　　　2. 5 are black and gray, two are black, gray and red, 3 are constructed from a variety of pastel colors

　　C. Method of administration
　　　　1. Blots presented in specific order that Rorschach prescribed.
　　　　2. Blots first shown to S in a performance or free association phase in which the S is asked to indicate what he sees in the blot.
　　　　3. Then clinician goes through the blots again to ascertain what aspects of the blot led to the response. This is the inquiry phase.
　　　　4. Then a phase in which the examiner tries to elicit from the S the usual or typical responses to each blot - testing the limits.
　　　　5. Test usually takes 45-55 minutes.

　　D. Scoring the Rorschach
　　　　1. Most common and well-developed system is the one from John Exner, starting in mid 1970's. In 1985 about 35% of practicing clinicians using Exner system.
　　　　2. All systems including Exner's, use roughly the same factors in scoring:
　　　　　　a. **Location**: What part of blot used: e.g. whole or part.
　　　　　　b. **Determinants**: forms, shading, color or apparent movement.
　　　　　　c. **Content**: human, animal or object.
　　　　　　d. **Original** or popular response: E.g. are perceptions bizarre, or usual for this blot?
　　　　　　e. **Form level**: Is the percept congruent with the characteristics of the blot.
　　　　　　f. **Relationships** between objects seen in several blots: E.g. aggression, submission to authority, danger, etc. Thematic similarities.

　　E. Data on reliability and validity of Exner system
　　　　1. **Exner, Armbruster & Viglione (1978)**: In nonpatient adult sample, used correlational analysis of 19 Rorschach variables and elements. Indicated that scores based on Exner categories are stable of 3-year period. 9 of 19 variables had test-retest correlations over .80.
　　　　2. For validity, unclear, but recent work more promising.
　　　　3. Notion of incremental validity: Clinicians complain that studies showing lack of validity do not reflect the way they actually use the instrument - in conjunction with clinical interviews and other information. May provide useful information in that context.

VIII. Objective tests: The Minnesota Multiphasic Personality I

  A. Introduction to Objective Tests
    1. No specific assumption about the appropriate perspective on personality
    2. Taps overt, conscious factors: feelings, attitudes, and personal characteristics. This report is taken as a true indication of those feelings, attitudes and characteristics.
    3. Are distinguished from projective tests by restricting the number of alternatives (this is the primary characteristic that differentiates objective from projective tests)
    4. Always have clear and standardized administration and scoring procedures.
    5. More likely than projective tests to have a clear, standardized and relatively objective and interpretation process.

IX. MMPI-2: Minnesota Multiphasic Personality Inventory

  A. Introduction and history
    1. Originally published in 1942 by Starke Hathaway (Ph.D.) and J. Charnley McKinley (M.D.) of Univ. of Minnesota Hospitals as a device to assist in diagnosis of psychiatric patients.
    2. Unlike previous tests in which content validity was used to choose items, authors used empirical validity to select items.
    3. Authors chose large number of items from variety of sources (psychiatric texts, existing tests, descriptions of psychiatric examinations).
    4. Chose items whose answers differentiated between one pathological group and others, and between than group and normals.
    5. MMPI-2 (1989) has 567 statements grouped into 10 clinical scales, each representing particular psychiatric problem. Subject indicates whether each statement applies to him or her by responding "YES", "NO", or "CAN'T SAY" to each.

  B. Scoring
    1. Item is scored as a plus on a scale if test-taker answers it in the same way as those in the clinical group. Test includes 3 validity scales.
      a. "I am very energetic."
      b. "People are out to get me."
      c. "I never have trouble falling asleep."
      d. "When I am bored, I often wind up getting into trouble."
      e. "I find it difficult to get out of the bed in the morning."
      f. "I am happy most of the time."
      g. "I hardly ever lose an argument."
      h. "I enjoy social gatherings just to be with people."
      i. "I believe I am a condemned person."
    2. Results (scored by hand or computer) presented as profile, connecting the scores on each scale. Raw scores transformed to **standard scores**, (T scores) with mean of 50

and standard deviation of 10. High scores are typically those over 70, though same score (eg 75) may mean different levels of abnormality on different scales.

C. **Clinical Scales**: Each scale has name indicating original clinical group it was designed to detect. But since high scores on single scale do not correlate well with specific disorder, clinicians generally use scale number rather than name.
   1. Scale 1 - (Hs) **Hypochondriasis**: High scorers described as cynical, critical, demanding and self-centered. Not good candidates for psychotherapy.
      a. "I do not tire quickly." (F)
      b. "I feel weak all over much of the time." (T)
      c. "I have very few headaches." (F)
   2. Scale 2 - (D) **Depression**: High scorers moody, shy, despondent, pessimistic, distressed.
      a. "My sleep is fitful and disturbed." (T)
      b. "I certainly feel useless at times." (T)
      c. "I brood a great deal." (T)
   3. Scale 3 - (Hy) **Hysteria**: High scorers outgoing, but repressed, naive, psychologically immature.
      a. "It takes a lot of argument to convince most people of the truth." (F)
      b. "I think most people would lie to get ahead." (F)
      c. "What others think of me does not bother me." (F)
   4. Scale 4 - (Pd) **Psychopathic deviate**: High scorers impulsive, hedonistic, antisocial; often have trouble with authority. Low scorers are conventional, conforming, moralistic.
      a. "I believe that my home life is as pleasant as that of most people I know." (F)
      b. "I have never been in trouble with the law." (F)
      c. "I have used alcohol excessively." (T)
   5. Scale 5 - (MF) **Masculinity-Femininity**: Measures identification with culturally conventional sex-typed interest patterns. (items are obvious and easy to fake) High males are sensitive, aesthetic, passive or feminine. High females are aggressive, rebellious, unrealistic. Highly educated males show up higher.
      a. "I would like to be a florist."
      b. "I like science."
      c. "I enjoy reading love stories."
   6. Scale 6 - (Pa) **Paranoia**: High scorers suspicious, aloof, shrewd, guarded, overly sensitive. Sometimes paranoid persons get very low score, since they are being defensive.
      a. "I am sure I get a raw deal from life." (T)
      b. "I believe I am being followed." (T)
      c. "I have no enemies who really wish to harm me." (F)
   7. Scale 7 - (Pt) **Psychasthenia**: Reflects chronic or trait anxiety, self-doubt, general dissatisfaction, agitated concern about self. High scorers tense, anxious, ruminative, preoccupied, obsessional, phobic, rigid. Feel inferior and inadequate.
      a. "Life is a strain for me much of the time." (T)
      b. "Almost every day something happens to frighten me." (T)
      c. "I have more trouble concentrating than others seem to have." (T)

8. Scale 8 - (Sc) **Schizophrenia**: Reflects feelings of being different, of isolation, bizarre or peculiar thought processes and perceptions, tendency to withdraw into fantasy. High scorers withdrawn shy, have unusual thoughts or ideas.
   a. "I have strange and peculiar thoughts." (T)
   b. "No one seems to understand me." (T)
   c. "I often feel as if things were not real." (T)
9. Scale 9 - (Ma) **Mania**: High scorers sociable, outgoing, optimistic, restless, impulsive; sometimes flighty, confused, disoriented.
   a. "Once a week or oftener I become very excited." (T)
   b. "I do not blame a person for taking advantage of someone who lays himself open to it." (T)
   c. "I have had periods of such great restlessness that I cannot sit long in a chair." (T)
10. Scale 0 -(Si) **Social Introversion-Extraversion**: High scorers are modest, withdrawn, inhibited. Low scorers outgoing, sociable, confident.
    a. "I find it hard to make talk when I meet new people." (T)
    b. "I like to be in a crowd who plays jokes on one another." (F)
    c. "I seem to make friends about as quickly as others do." (F)

D. **Validity scales**:
   1. In addition to clinical scales, test items are grouped on 8 validity scales, designed to detect individuals not answering appropriately or honestly. This could be due to:
      a. Inattention and carelessness
      b. Difficulty in reading or understanding the questions.
      c. Boredom, or lack of motivation (random or one-way answering)
      d. Deliberate attempts to look good or bad.
   2. Eight validity scales (three main ones):
      a. (L) **Lie scale**: 15 items selected to detect simple attempts to 'fake good' Contains items rationally chosen to reflect common weaknesses, and usually answered a certain way if the respondent is honest:
         (1) "Sometimes I get so mad I want to cry." (F)
         (2) "I always tell the truth." (T)
         (3) "I get angry sometimes." (F)
         (4) "I do not like everyone I know." (F)
         (5) "I do not read every editorial in the newspaper every day." (F)
      b. (K) **Defensive scale**: 30 items chosen to detect defensiveness more subtle than L scales. High score indicates defensiveness, low score may indicate excessive self-criticism or attempt to 'fake bad'. A high K score may invalidate test. Scores on most clinical scales are elevated by some constant fraction of the K-scale score. Includes items like:
         (1) "I think nearly everyone would tell a lie to keep out of trouble." (F)
         (2) "What others think does not bother me." (F)
         (3) "I certainly feel useless at times." (F)
         (4) "At times I feel like swearing." (F)
         (5) "At times I feel like smashing things." (F)
         (6) High scores usually indicate faking good, or a reluctance to admit psychopathology.
         (7) Hysterical patients (esp. conversion reaction) have an elevated K score.

c. (F) **Careless scale**: 60 items that are rarely agreed to by people; can be seen as reflecting the number of seriously psychopathological items endorsed. Designed to detect deviant or atypical ways of responding.
   (1) "My soul sometimes leaves my body." (T)
   (2) "Someone has control over my mind." (T)
   (3) "Everything tastes the same." (T)
   (4) "I see things, animals or people around me that others do not see." (T)
   (5) Several reasons for elevated F scores:
      (a) reading difficulty
      (b) deliberate attempt to look bad
      (c) psychotic processes
      (d) plea for help by patient exaggerating symptoms
      (e) in adolescents, defiance, hostility and negativism
d. (S) **Superlative Presentation Scale**, 50 items: Used in personnel testing to identify individuals who claim high moral values, few or no personal faults, no adjustment problems
e. (F(B)) **Infrequency-Back Scale**: 40 rarely endorsed items near the end of the MMPI-2. F scale items are all at the beginning (within first 370 items)
f. (F(P)) **Psychiatric Infrequency Scale**: To detect malingering of psychological symptoms.
g. (VRIN) **Variable Response Inconsistency Scale**: Used to detect random responding.
h. (TRIN) **True Response Inconsistency Scale**: Detects tendency to answer True, or to answer False, without regard to item content.

E. Interpreting MMPI results is done by looking at the overall profile, and/or at the 1, 2, or 3 scales with the highest score. A number of specific two-point and three-point profiles have been described.

F. Data on forensic reliability
  1. Test-retest reliability over short periods of time is good.
    a. Over a single day or less, about .80-.85 for normals and psychiatric patients (less for criminals)
    b. Over 1-2 weeks, .70-.80 for normals, higher for psychiatric cases, lower (.60-.70) for criminals
    c. Long-term test-retest reliability not too high: Over a year or more, .35-.45 for normals, .50-.60 for psychiatric patients, no data for criminals.

G. Data on forensic validity:
  1. The validity of the MMPI for making clinical judgements is not high, but is better than for most other single measures.
  2. Many studies show correlates between individual scale scores and non-test variables for a variety of populations. Studies also show correlates between configurations of two or more scales and non-test variables for an equal variety of populations.
  3. Rogers & Shuman do not support the use of the MMPI-2 in insanity evaluations

H. Criticisms of MMPI

1. Takes a long time to complete 60-90 minutes for normals, sometimes over 2 hours for others.
2. Some items found on several scales - as many as six, so ten clinical scales are highly intercorrelated.

X. Millon Multiaxial Clinical Inventory MCMI-III

   A. General Description and History
      1. Dates from the early 1980s
      2. Based on the theory of personality - and personality disorders - of Theodore Millon from the late 1960s:

   B. Test construction - like the MMPI
      1. **MCMI-II**I (1994): Has 24 clinical scales in four clusters:
         a. Personality style scales
         b. Severe personality scales
         c. Clinical syndrome scales
         d. Severe clinical syndrome scales

XI. MCMI-III: 175 T/F items taking about 30 minutes to complete; Designed to be used with emotionally disturbed patients

   A. 14 Personality Disorder Scales: Coordinate with DSM-IV Axis II disorders
      1. 11 **Moderate Personality Disorder** Scales:
         a. 1 - **Schizoid** (16 items): "Individuals are socially detached; prefer solitary activities; seem aloof, apathetic and distant with difficulties in forming and maintaining relationships." (Strack, 2008; p. 5)
         b. 2A - **Avoidant** (16 items): "Individuals are socially anxious due to perceived expectations of rejection." (Strack, 2008; p. 5)
         c. 2B - **Depressive** (15 items): "Individuals are downcast and gloomy, even in the absence of a clinical depression." (Strack, 2008; p. 5)
         d. 3 - **Dependent** (16 items): "Individuals are passive, submissive, and feel inadequate. They generally lack autonomy and initiative." (Strack, 2008; p. 5)
         e. 4 - **Histrionic** (17 items): "Individuals are gregarious, with a strong need to be at the center of attention. They can be highly manipulative." (Strack, 2008; p. 5)
         f. 5 - **Narcissistic** (24 items): "Individuals are self-centered, exploitive, arrogant and egotistical." (Strack, 2008; p. 5)
         g. 6A - **Antisocial** (17 items): "Individuals are irresponsible, vengeful, engage in criminal behavior, and are strongly independent." (Strack, 2008; p. 5)
         h. 6B - **Aggressive** (Sadistic) (20 items):"Individuals are controlling and abusive; they enjoy humiliating others." (Strack, 2008; p. 5)

    i. 7 - **Compulsive** (17 items): "Individuals are orderly, organized, efficient, and perfectionistic. They engage in these behaviors to avoid chastisement from authorities." (Strack, 2008; p. 5)

    j. 8A - **Passive-Aggressive** (Negativistic) (16 items): "Individuals are disgruntled, argumentative, petulant, oppositional, negativistic; they keep others on edge."(Strack, 2008; p. 5)

    k. 8B - **Self-Defeating** (Masochistic) (15 items): "Individuals seem to engage in behaviors that result in people taking advantage of and abusing them. They act like a martyr [sic] and are self-sacrificing." (Strack, 2008; p. 5-6)

2. 3 **Severe personality Pathology** Scales
   a. S - **Schizotypal** (16 items): "Individuals seem 'spacey', self-absorbed, idiosyncratic, eccentric and cognitively confused." (Strack, 2008; p. 6)
   b. C - **Borderline** (16 items): "Individuals display a labile affect and erratic behavior. They are emotionally intense, often dissatisfied and depressed, and may become self-destructive." (Strack, 2008; p. 6)
   c. P - **Paranoid** (17 items): "Individuals are rigid and defensive. They hold delusions of influence and persecution. They are mistrusting and may become angry and belligerent." (Strack, 2008; p. 6)

3. 10 Clinical Syndrome Scales:

4. **Moderate Syndrome Scales**
   a. A - **Anxiety** (14 items): "Individuals are anxious, tense, apprehensive, and physiologically over-aroused." (Strack, 2008; p. 6)
   b. H - **Somatoform** (12 items): "Individuals are preoccupied with vague physical problems with no known organic cause. They tend to be hypochondriacal and somaticizing." (Strack, 2008; p. 6)
   c. N - **Bipolar: Manic** (13 items): "Individuals have excessive energy and are overactive, impulsive, unable to sleep, and are manic." (Strack, 2008; p. 6)
   d. D - **Dysthymia** (14 items): "Individuals are able to maintain day-to-day functioning but are depressed, pessimistic, and dysphoric. They have low self-esteem and feel inadequate." (Strack, 2008; p. 6)
   e. B - **Alcohol Dependence/Abuse** (15 items): "Individuals admit to serious problems with alcohol and/or endorse personality traits often associated with abusing alcohol." (Strack, 2008; p. 6)
   f. T - **Drug Dependence/Abuse** (14 items): "Individuals admit to serious problems with drugs and/or endorse personality traits often associated with abusing drugs." (Strack, 2008; p. 6)
   g. R - **Post-Traumatic Stress Disorder** (16 items) - not included on previous versions of MCMI: "Individuals report intrusive or unwanted memories and/or nightmares of a disturbing or traumatic event; they may have flashbacks." (Strack, 2008; p. 6)

5. **Severe Syndrome Scales**
   a. SS - **Thought Disorder** (17 items): "Individuals experience thought disorder of psychotic proportions; they often hallucinations and delusions." (Strack, 2008; p. 6)
   b. CC - **Major Depression** (17 items): "Individuals are severely depressed to the extent that they are unable to function in day-to-day activities. They have vegetative signs of clinical depression (poor appetite and sleep, low energy, loss of

interests) and feel hopeless and helpless." (Strack, 2008; p. 6)
    c. PP - **Delusional Disorder** (Psychotic Delusion) (13 items): "Individuals are acutely paranoid with delusions and irrational thinking. They may become belligerent and act out their delusions." (Strack, 2008; p. 6)

B. 3 Modifying Indices
    1. X - **Disclosure**: "Scale X measures the amount of disclosure and willingness to admit to symptoms and problems." (Strack, 2008; p. 5)
        a. Calculated by adding up raw scores from 10 basic personality scales
        b. Low scores = secretive and defensive
        c. High scores = openly frank and revealing
        d. Scores <34 or >178 invalidate the test
    2. Y - **Desirability**: "Scale Y measures examinee's tendency to answer items that make one look very favorable and without problems." (Strack, 2008; p. 5)
        a. High scores = denying problems; may be expected in custodial situations or job retention evaluations; do not invalidate the profile
        b. Low scores not interpreted on this scale
    3. Z - **Debasement**: "Scale Z assesses examinee's tendency to answer items by accentuating, highlighting, and exaggerating problems and symptoms." (Strack, 2008; p. 5)
        a. May be cry for help, emotional turmoil, or symptom exaggeration for personal gain
        b. High scores do not invalidate the profile

C. V **Validity Index**: "Two [actually three] items measuring highly improbably events designed to detect random responding and confusion" (Strack, 2008; p. 5)
    1. If two of the three items are endorsed, the test is invalidated

D. Scoring and Interpretation
    1. ~~Using table in the manual, raw scores on each scale converted to Base Rate (BR) scores~~
        a. BR scores range from 0 to 115
        b. BR of 60 represents median raw score for all patients
    2. BR of 75 set at minimum score for patients who met DSM-III criteria for particular disorder or condition
        a. BR scores from 75-84 indicate clinically significant personality style or syndrome
        b. BR >85 indicate prominent personality style or syndrome: Set at raw score that indicated personality or syndrome as the person's primary condition

E. Additional versions or tests in addition to MCMI:
    1. MCMI-III Corrections Report
        a. **Inmate correctional sample**: "Uniquely based on corrections norms, the MCMI-III Corrections Report presents targeted information to help psychologists and corrections professionals make security, management, and treatment decisions faster and more accurately. In addition to providing all of the clinical information to which MCMI-III test users are accustomed, the Corrections Report is distinguished by the inclusion of a one-page Correctional Summary of likely needs and behaviors relevant to correctional settings. Now, this highly useful summary

has been further strengthened by the addition of empirically based statements that classify an offender's probable need as High, Medium, or Low in three critical areas:
- (1) Mental health intervention
- (2) Substance abuse treatment
- (3) Anger management services.

    b. "Developed through a recent study that involved more than 10,000 offenders, these statements are designed to help inform staff of likely offender behaviors to help support crucial programming and placement decisions. In addition, the summary provides clinically based statements on six issues of most concern in corrections settings:
- (1) Reaction to Authority
- (2) Escape Risk
- (3) Disposition to Malinger
- (4) Response to Crowding / Isolation
- (5) Amenability to Treatment / Rehabilitation
- (6) Suicidal Tendencies

F. Issues with the MCMI-III
1. High test-retest reliability, especially on personality scales, less so on clinical syndrome scales
2. Normative sample small and not representative of minority groups
3. High degree of scale overlap: Because of high number of scales and relatively low number of items, correlations between scales range from .40 to .85
4. The vast majority of items scored as "True", so vulnerable to patients with acquiescent response bias
5. Weak in assessing major psychotic disorders or patients with minor personality pathology
6. Some scales (Histrionic, Narcissistic, Compulsive) have trouble diagnosing patients with corresponding disorders; seem better at revealing personality styles.
7. Many scales show poor convergent validity with other psychiatric rating instruments
8. Millon's model of personality disorders not strongly validated - little research
9. MCMI-III too new to have generated many research studies to support it - and this is important because large changes between versions I, II, and III mean that many people see them as different instruments
10. Richard Rogers strongly opposed to it in insanity evaluations: "*The MCMI-III should not be used in insanity evaluations based on its diagnostic invalidity. For Axis II disorders, elevation of a designate scale as evidence of the corresponding disorder is likely to be wrong 82% of the time. The numbers are not much better for Axis I disorders in which a designated scale is wrong 69% of the time. These levels of inaccuracy are below the threshold for admissibility of relevant testimony.*" (Rogers & Shuman, p. 202)

XII. Rogers Criminal Responsibility Assessment Scale (R-CRAS)

   A. General Description and History
      1. Published in 1984. Designed to *"quantify essential psychological and situational variables at the time of the crime and to implement criterion-based decision models for criminal responsibility."* (Rogers, 1984, p. 1)
      2. Rogers worked with a study group of four forensic psychologists and three forensic psychiatrists to conceptualize and operationalize key constructs to be investigated in assessing criminal responsibility. (Rogers & Shuman, p, 230)
      3. Consists of 25-30 variables (depending on who's counting) that are typically rated on a 5-point or 6-point scale:
         a. 0 – "No Information"
         b. 1 = "Not Present"
         c. 2 = "Clinically insignificant"
         d. 3-6 = increasing gradations of clinically relevant symptoms
      4. Assesses criminal responsibility using the American Law Institute standard: Insane *"if at the time of his conduct as a result of mental disease or defect he lacks substantial capacity either to appreciate the criminality (wrongfulness) of his conduct or to conform his conduct to the requirements of law."*

   B. A number of different specific variables:
      1. Malingering (2 scales)
      2. Organic mental disorder
      3. Mental retardation
      4. Amnesia
      5. Anxiety
      6. Bizarre behavior
      7. Delusions
      8. Depressed or elevated mood
      9. Affective Disorder
      10. Hallucinations
      11. Thought Disturbance ("evidence of formal thought disorder")
      12. Language Disturbance ("patient's level of verbal coherence at the time of the crime")
      13. Awareness of Criminality ("awareness of criminality during the commission of the alleged crime")
      14. Evidence of Planning - ("Planning and preparation for the alleged crime")

   C. Six summary psychological criteria based on these variables
      1. A1 - presence of malingering
      2. A2 - presence of organicity
      3. A3 - presence of major psychiatric disorder
      4. A4 - Ability to comprehend criminality of behavior ("definite loss of cognitive controls")
      5. A5 - Loss of behavioral control
      6. A6 was loss of control due to
         a. Organic disturbance
         b. Psychiatric disturbance

D. Reliability:
 1. Rogers (1984): Reliability, as assessed by separate R-CRAS assessments about 3 weeks apart fairly high:
    a. For 23 different variables, 12 had >= .60; 8 had r >.40 -<.60; and 3 had r <.40
    b. Expert agreement:
       (1) Malingering: 85%
       (2) Major mental disorder: 88%
       (3) Cognitively aware of crime: 87%
       (4) Able to control criminal behavior: 89%
    c. But 'experts' doing the same thing that the R-CRAS does; interviewing and assessing

E. Validity:
 1. Rogers notes the problems here since validity in the most general sense will mean agreeing with an evolving and subjective standard of insanity.
 2. Mentions three kinds of validity:
    a. **Substantive validity** (content validity): Do the items in the scale address elements of the construct. Notes that no critics take issue with the substantive validity of the R-CRAS.
    b. **Structural validity**: Do individuals judged insane according to R-CRAS meet common criteria of, or characteristics of, insanity? Yes:
       (1) Relative absence of malingering
       (2) More severe psychological impairment
       (3) Greater loss of cognitive and/or behavioral control, than those judged criminally responsible.
       (4) But **Melton et al (1997)** argue that this is "trivial and tautological" since these same characteristics are the basis on which individuals would be assessed as being not criminally responsible.
    c. **External validity** (criterion validity): How do the judgments based on R-CRAS match legal decisions? Rogers (1984) report 85% agreement:
       (1) 73.3% for those assessed as insane
       (2) 95.2% for those judged as sane
       (3) But again, the R-CRAS assessment would be part of the court testimony, so the verdict is not independent of the assessment.

F. R-CRAS reflects relative importance assigned by examiners to the first-order elements of insanity-decision (e.g., the presence and relevance of psychopathology to mental state at the time of the offense).
 1. These elements fairly abstract psychological and legal terms (e.g., "delusions at the time of alleged crime") and do not necessarily represent the cues that are actually utilized by professional examiners in making their decisions.
 2. This is a major issue of contention between Rogers, Melton et al. (1997), and Golding (1992) in the evaluation of R-CRAS. Rogers believes it important to quantify issue, although unfair to assume he does not recognize value of more qualitative data (see **Rogers & Ewing, 1992**).
 3. Melton and Golding agree that quantification is essentially illusory at this stage in the

development of evaluations of criminal responsibility. Both groups of authors agree, in large measure, on the domain of conceptual elements to be addressed.

G. Factor analysis of the R-CRAS items results in three factors:
1. **Bizarre behavior**, **high activity**, and **high anxiety** - that do not mirror the five scales (see Borum, 2003).
2. Final judgments with the R-CRAS also show reasonable levels of agreement between examiners and triers of fact (96% with respect to sanity with lower levels of agreement on insanity [70%]; **Rogers, Cavanaugh, Seeman & Harris, 1984**; see **Rogers & Shuman, 2000** for a summary).
3. Unfortunately, all studies in this area appear to use criterion contaminated groups in that the examination process is part of the judicial/criterial determination.

References

Alheidt, P. The effect of reading ability on Rorschach performance. *Journal of Personality Assessment*, 1980, 44, 3-10.

Aronow, E. & Reznikoff, M. A. *Rorschach Introduction: Content and Perceptual Approaches*. New York: Grune and Stratton, 1983.

Bernard, G.W., Thompson, J.W., Freeman, W.C., Robbins, L. Gies, D. & Hankins, G.C. Competency to Stand Trial: Description and Initial assessment of a new computer-assisted assessment tool (CADCOMP. 1991, *Bulletin of the American Academy of Psychiatry*. 1991, 19(4), 367-371.

Borum, R. & Grisso, T. Psychological test use in criminal forensic evaluations. *Professional Psychology: Research and Practice*, 1995, 26, 465-473.

Borum, R. (2003). Not guilty by reason of insanity. In T. Grisso (Ed.), *Evaluating competencies (2nd ed.)*. New York: Kluwer/Plenum.

Breidenbaugh, B., Brozovich, R. & Matheson, L. The had test and other aggression indicators in emotionally disturbed children. *Journal of Personality Assessment*, 1974, 38, 332-334.

Choca, James P. *Interpretive Guide to the Millon Clinical Multiaxial Inventory (3rd Edition)*. Washington, D.C.: APA, 2004.

Exner, J., Thomas, E.A. & Mason, B. Children's Rorschachs: Description and prediction. *Journal of Personality Assessment*, 1985, 49, 13-20.

Gazzaniga, M. *Mind Matters*. Boston: Houghton Mifflin, 1988. P. 233.

Golding, S. (1992). The adjudication of criminal responsibility: A review of theory and research. In D. Kagehiro & W. Laufer (Eds.), *Handbook of Psychology and Law*. New York: Springer-Verlag.

Hammer, Emanual F. Projective drawings. In A.I. Rabin (Ed.) *Assessment with Projective Techniques*. New York: Springer Publishing, 1981.

Hammer, M. & Kaplan, A. The reliability of children's human figure drawings. *Journal of Clinical Psychology*, 1966, 22, 316-319.

Hathaway, S. R. & McKinley, J. C. *Minnesota Multiphasic Personality Inventory*. Minneapolis: University of Minnesota Press, 1942.

Kagan, J. The measurement of overt aggression from fantasy. *Journal of Abnormal and Social Psychology*, 1956, 52, 390-393.

Klein, R.G. Questioning the clinical usefulness of projective psychological tests for children. *Developmental and Behavioral Pediatrics*, 1986, 7, 378-382.

Kline, P. *Psychological Testing*. London: Malaby Press, 1976.

Lifshitz, M. Social differentiation and organization of the Rorschach in fatherless and two-parented children. *Journal of Clinical Psychology*, 1975, 31, 126-130.

Lubin, B., Larsen, R.M., Matarazzo, J.D. & Seever, M. Psychological test usage patterns in five professional settings. *American Psychologist*, 1985, 40, 857-861

Morgan, C.D. & Murray, H.A. A method for investigating fantasies: The Thematic Apperception Test. *Archives of Neurology and Psychiatry*, 34, 289-306.

Lubin, B., Larsen, R.M., Matarazzo, J.D. & Seever, M. Patterns of psychological test usage in the United States: 1935-1982. *American Psychologist*, 1984, 39, 451-454.

Lubin, B., Larsen, R.M., Matarazzo, J.D. & Seever, M. Psychological test usage patterns in five professional settings. *American Psychologist*, 1985, 40, 857-861

First, M.B, Spitzer, R.L, Williams, J.B.W. & Gibbon, M. *Structured Clinical Interview of DSM-IV Disorders*. Washington, D.C.: APA, 1997.

Lifshitz, M. Social differentiation and organization of the Rorschach in fatherless and two-parented children. *Journal of Clinical Psychology*, 1975, 31, 126-130.

Exner, J., Thomas, E.A. & Mason, B. Children's Rorschachs: Description and prediction. *Journal of Personality Assessment*, 1985, 49, 13-20.

Morgan, C.D. & Murray, H.A. A method for investigating fantasies: The Thematic Apperception Test. *Archives of Neurology and Psychiatry*, 34, 289-306.

Kagan, J. The measurement of overt aggression from fantasy. *Journal of Abnormal and Social Psychology*, 1956, 52, 390-393.

Hammer, Emanual F. Projective drawings. In A.I. Rabin (Ed.) *Assessment with Projective Techniques*. New York: Springer Publishing, 1981.

Breidenbaugh, B., Brozovich, R. & Matheson, L. The had test and other aggression indicators in emotionally disturbed children. *Journal of Personality Assessment*, 1974, 38, 332-334.

Hammer, M. & Kaplan, A. The reliability of children's human figure drawings. *Journal of Clinical Psychology*, 1966, 22, 316-319.

Holub, R.J. *Forensic psychological testing: A survey of practices and beliefs*. Unpublished manuscript, Minnesota School of Professional Psychology, Bloomington, MN. 1992.

Klein, R.G. Questioning the clinical usefulness of projective psychological tests for children. *Developmental and Behavioral Pediatrics*, 1986, 7, 378-382.

Kline, P. *Psychological Testing*. London: Malaby Press, 1976.

Melton, G. B., Petrila, J., Poythress, N. G., & Slogobin, C. (1997). *Psychological evaluations for the courts: A handbook for mental health professionals and lawyers (2nd ed.)*. New York: Guilford.

Nicholson, R.A. & Kugler, K. Competent and incompetent criminal defendants: A quantitative review of comparative research. *Psychological Bulletin*, 1991, 109, 355-370.

Roesch, R., Eaves, D., Sollner, R., Normandin, M. & Glackman, W. Evaluating fitness to stand trial: A comparative analysis of fit and unfit defendants. *International Journal of Law and Psychiatry*, 1981, 4, 145-157.

Roesch, Ronald., Zapf, Patricia.A., Eaves, D. & Webster, C.D. *The Fitness Interview Test (rev. ed.)* Available from Mental Health, Law, and Policy Institute, Simon Fraser University, 1998

Rogers, R. *Rogers Criminal Responsibility Assessment Scales*. Psychological Assessment Resources. Odessa, FLA, 1984.

Rogers, R., Salekin, R.T. & Sewell, K.W. Validation of the Millon Clinical Multiaxial Inventory for Axis II disorders: Does it meet the Daubert standard? *Law and Human Behavior*, 1999, 23, 425-443.

Rogers, R., Salekin, R.T. & Sewell, K.W. The MCMI and the Daubert standard: Separating rhetoric from reality. *Law and Human Behavior*, 2000, 24, 501-506.

Rogers, R. & Sewell, K.W. The R-CRAS and insanity evaluations: A re-examination of construct validity. *Behavioral Sciences and the Law*, 1999, 17, 181-194.

Rogers, R., Wasyliw, O. E., & Cavanaugh, J. L. (1984). Evaluating insanity: A study of construct validity. Law and Human Behavior, 8, 293-303.

Rogers, R. & Shuman, D.W. *Conducting insanity evaluations* (2nd edition). New York: Guilford Press, 2000.

Spitzer, R.L. & Endicott, J. *Schedule of Affective Disorders and Schizophrenia* (3rd Ed.). New York: Biometrics Research, 1978.

Ustad, K.L., Rogers, R., Sewell, K.W. & Guarnaccia, C.A. Restoration of Competency to Stand Trial: Assessment with the Georgia Court Competency Test and the Competency Screening Test. *Law and Human Behavior*, 1996, 20(2), 131-146..

Webster, C.D., Menzies, R.S., Butler, B.T. & Turner, R.E. Forensic psychiatric assessment in selected Canadian cities. *Canadian Journal of Psychiatry*, 1982, 27, 455-462.

Zapf, P.A. & Roesch, R. Assessing fitness to stand trial: a comparison of institution-based evaluations and a brief screening interview. *Canadian Journal of Community Mental Health*, 1997, 16, 53–66.

Zapf, P.A. & Roesch, R. Fitness to stand trial: Characteristics of remands since the 1992 criminal code amendments. *Canadian Journal of Psychiatry*, 1998, 43, 287-293.

Zapf, P.A., Roesch R, & Viljoen Jodi L. The utility of the Fitness Interview Test for assessing fitness to stand trial. *Canadian Journal of Psychiatry*, 2001, 46, 426–432.

Zapf P.A. & Roesch R. A comparison of MacCAT-CA and the FIT for making determinations of competency to stand trial. *International Journal of Law and Psychiatry*, 2001, 24, 81–92.

Violence Risk Assessment

I. Introduction

    A. Important for:
        1. sentencing decision with respect to how long before parole
        2. assessing treatment to be offered/mandated while in custody
        3. assessing parole eligibility

    B. Also may be interested in risk factors for violence, and how they may be reduced (risk management)

    C. Includes both sexual aggression and non-sexual violence

II. History in U.S.

    A. *Baxtrom v. Herold* (U.S. Supreme Court, 1966)
        1. Baxtrom a N.Y. state prisoner, diagnosed as mentally disordered and transferred to hospital for criminally insane just before expiration of his prison term.
        2. Was held past his prison term without a finding of current dangerousness, which Court rules a violation of his due process rights.
        3. As a result of decision, 967 offenders deemed to be too dangerous to be released from maximum security hospitals in N.Y. were released to lower-security civil hospitals. Actual rates of reoffending were low:
            a. Only 20% of offenders were subsequently violent
            b. 18% were discharged into the community within a year of admission
            c. Less than 1% subsequently returned to secure hospitals
            d. Over 4.5 year period, more than 50% of patients discharged to the community, and fewer than 3% returned to secure hospitals. (**Monahan & Steadman 2001; Steadman & Cocozza, 1974**)

    B. **Thorberry & Jacoby (1979)** Studied 586 patients released from Pennsylvania's Fairview institution as a result of *Dixon v Pennsylvania* (1971)
        1. Three-year recidivism rate was 23.7%
        2. Only 15% rearrested or returned for violent offense during the 4-year follow-up period

    C. **Quinsey & Ambtman (1979)**: Experienced psychiatrists predictions of violence no better than that of school teachers, and neither group made much use of psychiatric assessment data.

D. Prediction research vanished in the 1980s after **Monahan's (1981)** *Clinical Prediction of Violent Behavior* argued that "psychiatrists and psychologists are accurate in no more than one out of three predictions of violent behavior" (p.47)

E. U.S, supreme Court decision in *Barefoot v. Estelle* (1983): Plaintiff objected to use of psychiatrists' testimony regarding his risk of future dangerousness, arguing the psychiatrists were not competent to make such predictions.
   1. American Psychiatric Association filed *amicus curiae* brief supporting plaintiff's contention, and concluding that psychiatrists have no expertise in predicting future violence, and that laypersons could do as well.
   2. Brief argued that psychiatrists more often wrong than right, and tended to overpredict violence (Monahan, 1981; Steadman & Coccoza, 1974; Thornberry & Jacoby, 1979)
   3. Court ignored brief, arguing that "we are [not] convinced that the view of the APA should be converted into a constitutional rule barring an entire category of expert testimony." (*Barefoot v. Estelle*, 1983, p. 3387)
   4. Court: "neither petitioner nor the Association suggests that psychiatrists are always wrong with respect to future dangerousness, only most of the time."

F. What are the blind spots in clinical assessment of violence risk? (Monahan, 1981)

G. So given that such assessments must now occur, the focus of research in the area turned to the question of how they could be made more accurate.

III. Models of Risk Assessment

A. Actuarial methods (see **Quinsey et al, 2006**): Remover the inaccurate clinician from the process and rely on statistical data.

B. Increase clinician's accuracy by standardizing clinical evaluation process

C. Problems noted with traditional methods of calculating predictive accuracy within standard 2 x 2 predictive matrix:
   1. PPP (Positive Predictive Power); NPP (Negative Predictive Power); Sensitivity ; Specificity
      a. Both positive and negative predictive power are about the accuracy of what the test says.
      b. Positive Predictive Pwer (PPP) is the proportion of people that the test identifies as having some condition or risk, or belonging to some category, that actually do. So it is calculated as Hits divided by Hits and False alarms: (H/(H + FA)
      c. Negative Predictive Power (NPP) is the proportion of people that the test identifies as NOT having some condition or risk, or belonging to some category, that actually do NOT. So it is calculated as Correct Rejections (R) divided by R plus Misses: R/(R + M).

2. All are influenced by base rates, and perform best at 50% base rate.
3. See: http://www.medcalc.be/calc/diagnostic_test.php for online calculation of 2 x 2 statistics

D. ROC curves from signal detection theory now used more:
1. X-axis = false alarms
2. Y-axis = hits
3. Is the ratio of Sensitivity to (1- Specificity)
4. "Discrete classifiers, such as decision tree or rule set, yield numerical values or binary label. When a set is given to such classifiers, the result is a single point in the ROC space. For other classifiers, such as naive Bayesian and neural network, they produce probability values representing the degree to which class the instance belongs to. For these methods, setting a threshold value will determine a point in the ROC space. For instance, if probability values below or equal to a threshold value of 0.8 are sent to the positive class, and other values are assigned to the negative class, then a confusion matrix can be calculated. Plotting the ROC point for each possible threshold value results in a curve." (Wikipedia; http://en.wikipedia.org/wiki/ROC_curve)
5. Can define Area Under the Curve (AUC), which ranges from 1.00 (perfectly accurate predictions) to 0.00 (perfectly wrong predictions), and where .50 is chance prediction.
6. "The AUC is equal to the probability that a classifier will rank a randomly chosen positive instance higher than a randomly chosen negative one. It can be shown that the area under the ROC curve is equivalent to the Mann-Whitney U, which tests for the median difference between scores obtained in the two groups considered if the groups are of continuous data. It is also equivalent to the Wilcoxon test of ranks." (Wikipedia; http://en.wikipedia.org/wiki/ROC_curve)
7. For violence prediction, AUC defined as probability that randomly selected violent person will score higher on the measure than randomly chosen non-violent person.
8. AUCs of .70 and above considered large - though obviously not perfect!
9. Estimates from ROC analysis limited in their applicability to individual cases.

IV. Actuarial Prediction

A. In review of literature, Garb & Boyle (2003) show that unstructured clinical judgment prone to error in all psychological decision making.

B. **Quinsey, Pruesse & Fernley (1975)**: Followed 60 involuntary psychiatric patients released into the community:
1. 30% rearrested during 39-month period
2. Constructed scale assigning 1 point for each variable that discriminated between success (no violence) and failure:
   a. Diagnosis of a personality disorder
   b. Age under 31 at discharge
   c. Less than 5 years in psychiatric hospital

d. Admission offense not against people
e. Separation from parent before age 16
3. Scale classified patients with 78% accuracy

C. **Preusse & Quinsey (1977)**: Tried to validate above scale with 206 patients released from maximum security hospital. Accuracy dropped from 78% to 65% (common in cross-validation studies.)

D. Same colleagues developed Violence Risk Appraisal Guide (VRAG: **Harris, Rice & Quinsey, 1993**)
1. Designed to predict long-term violence risk among individuals with prior violent episodes.
2. Combined validation sample:
   a. 685 violent and sexually violent offenders at Oak Ridge, 618 of whom had chance to recidivate
   b. Included insanity acquittals and matched sample who had spent at least one day in Oak Ridge between 1975 and 1978
   c. Note that all were psychiatric patients, though authors argue that VRAG would work as well on non-psychiatric cases
3. Defined outcome measure as any criminal charge, or event that would have lead to criminal charge if committed in the community. Cases ranged from assault to homicide.
4. Collected 50 predictor variables on their sample, based on known relation to violence, or authors' interest in knowing such relationship.
   a. Childhood history:
      (1) highest school grade
      (2) elementary school maladjustment
      (3) teen alcohol abuse score
      (4) socioeconomic status
      (5) childhood aggression
      (6) behavior problems
      (7) suspended or expelled
      (8) arrested under age 16
      (9) separated from parents under age 16
      (10) parental crime
      (11) parental psychiatric history
      (12) parental alcoholism
   b. Adult adjustment:
      (1) longest employment (in months)
      (2) admissions to corrections
      (3) psychiatric admissions
      (4) alcohol abuse score
      (5) impulsivity score
      (6) property offense history
      (7) violent offense history
      (8) never married
      (9) previous violent offense

            (10) ever fired
            (11) escaped from an institution
            (12) failure on prior conditional release
            (13) lived alone
    c. Index offense:
            (1) age at index offense
            (2) victim injury
            (3) seriousness of index offense
            (4) violent offense
            (5) victim knew offender
            (6) female victim
            (7) weapon used
            (8) sexual motive
            (9) alcohol involved
    d. Assessment results:
            (1) IQ
            (2) Level of Supervision Inventory
            (3) Psychopathy Checklist
            (4) Elevation on MMPI scale 4 (Pd)
            (5) DSM-III Schizophrenia
            (6) DSM-III Personality Disorder
            (7) Procriminal values
            (8) Attitude unfavorable to convention
5. Items included for their empirical relation to prediction (like the MMPI), not for their face validity in the issue. Twelve variables chosen (with correlation with violent recidivism):
    a. Elementary school maladjustment (.31)
    b. Separated from parents under age 16 (.25)
    c. Alcohol abuse history (.13)
    d. Property offense history (.20)
    e. Never married (.18)
    f. Failure on prior conditional release (.24)
    g. Age at index offense (-.26)
    h. Victim injury in index offense (-.16)
    i. Female victim in index offense (-.11)
    j. Psychopathy Checklist (.34)
    k. DSM-III Schizophrenia (-.17)
    l. DSM-III Personality Disorder (.26)
6. Each item scored according to weighting procedure
7. Validity:
    a. **Harris et al (1993)**:
            (1) Found correlation of .44 between total VRAG scores and violent recidivism.
            (2) AUC in construction sample was .76, so randomly drawing violent and non-violent patient from sample would result in violent patient getting higher score 76% of the time.
    b. **Rice & Harris (1997)**: Cross-validated VRAG with sample of 159 sex offenders

who were not part of the construction sample. Looked at incidents of violence and sexual violence after release.
- (1) Correlation of .47 between VRAG scores and violent recidivism
- (2) Only .20 correlation between VRAG and sexual recidivism

c. Many researchers study VRAG alone:
- (1) Harris, Rice & Camilleri (2004)
- (2) Hilton & Simmons (2001)
- (3) Hilton, Harris & Rice (2001)
- (4) Loza, Villanueve & Loza-Fanous (2002)

d. Others compare VRAG with other measures predicting violent recidivism:
- (1) Barbareee, Seto, Langton & Peacock (2001)
- (2) Glover et al (2002)
- (3) **Kroner & Mills (2001)**: Compared predictive abilities for institutional misconduct and recidivism in violent, non-sexual offenders. Compared VRAG, HCR-20, PCL-R, Level of Service Inventory - Revised, and Lifestyle Criminality Screening Form.
  - (a) Found no overall differences in predictive accuracy among the five.
  - (b) Trend toward greater accuracy in VRAG for predicting minor disciplinary infractions and non-violent recidivism.
- (4) **Harris et al (2003.)**: Studied four Canadian forensic samples and compared VRAG to Sex Offender Risk Appraisal Guide (SORAG); Rapid Risk Assessment for Sex Offender Recidivism (RRASOR); and Static-99.
  - (a) general trend indicated superiority for VRAG and SORAG.
  - (b) across samples, AUC for VRAG was .73 for violent recidivism, for sexual recidivism it was .65

e. Jackson argues that VRAG is best validated actuarial risk assessment tool. Notes that Loza et al (2002) caution that performance of VRAG will be lower when base rate of violence is significantly different from that in the VRAG validation sample.

E. Sexual Offending Risk Appraisal Guide (SORAG) (Quinsey, Harris, Rice & Cormier, 1998)
1. Developed as an extension of VRAG to estimate risk of violence among past sexual offenders.
2. Includes 10 items from VRAG, but adds 4 items specific to sex offending. Because of overlap in items, test scores are highly correlated.
3. Good accuracy in predicting violent recidivism, and slightly less so predicting sexual recidivism.

F. Critique of the actuarial approach - though it seems to work
1. 'Violence' means different things to different instruments, and sexual offending can include non-contact offenses such as voyeurism and exhibitionism which are much less of an issue than assault and rape.
2. Actuarial methods give risk estimates for groups, not for individuals.
3. Actuarial methods focus on static, long-term risk levels; would like to know more about factors that change the momentary risk from day to day or week to week - especially for treatment or amelioration of risk.

G. Iterative Classification Tree (ICT) Model: Developed by **Steadman et al (2000)** in response to criticisms of the actuarial approach.
   1. Creation:
      a. Based on data from the MacArthur violence risk assessment study (Monahan et al, 2001) designed to assess risk of violence among released civil (not criminal) psychiatric patients.
      b. Released patients followed in the community for 5 months, using self report data, collateral reports from those in the community who knew the patients, and arrest records.
      c. Data led to examination of relationship between 134 possible risk factors and the outcome variable.
      d. Then started with variable most highly predictive of violence and divided sample into two groups on the basis of whether they possessed that risk factor or not.
      e. Then took those groups and partitioned them by additional risk factors.
      f. Then took those left unclassified as high- or low risk, pooled them, and started the process again (hence the 'iteration')
   2. Results:
      a. 76.6% of all participants were classified as either low- or high-risk for violence
      b. AUC was .82, which is very good indeed.
   3. **Silver et al (2000)**: Found that logistic regression was equal to the ICT model using the same data as Steadman et al (2000).
   4. **Monahan et al (2005)**: Used ICT with 700 civil psychiatric patients from three hospitals to predict community violence in 5 months post-discharge.
      a. 76% of patients were accurately classified as high- or low-risk, and AUC was .70
   5. New and not widely used by clinicians because it requires computer technology (i.e, programs) not readily available.

V. Structured clinical judgment

   A. Introduction
      1. In the risk assessment literature 'clinical judgment' usually means unstructured clinical judgment.
      2. Still no evidence that unstructured clinical judgment is any good - certainly not as good as actuarial assessment.
      3. But what about structured clinical judgment?
      4. **Douglas, Ogloff & Hart (2003)**: Structured professional judgment involved *"a clinician reviewing all relevant data sources for the presence of specified static and dynamic risk factors and then making a structured final risk judgment"* (Jackson, p. 165)
      5. Structured approached:
         a. Standardize how evaluations are conducted
         b. Standardize how variables are weighted
         c. But judgments are not made strictly on the basis of weighted criteria; clinical

judgment may modify that.

B. Two main vehicles for structured clinical assessment:
1. **Historical/Clinical Risk Management-20** (HCR-20)
2. **Sexual Violence Risk-20** (SVR-20)
3. Newer: **Level of Service Inventory-Revised (LSI-R)**

VI. Historical/Clinical Risk Management (HCR-20)

A. Developed by Christopher Webster, Kevin Douglas, and Stephen Hart, all of Simon Fraser University, and Derek Eaves of the B.C. forensic Psychiatric Services Commission.

B. 20 variables organized into three scales
1. **Historical scale**: 10 risk variables, mostly stable and static. "*historical data should anchor risk assessments. In general clinical settings, predictions based on a relatively small number of historical variables generally are as accurate, or more accurate, than those based on detailed information...; the same appears to be true in forensic settings.*" 'historical' "*reflects the temporal stability of these items, which tend to be static in nature. It does not imply that they are all unchangeable, ascriptive features of individuals (e.g., demographic variables) or past events.*" (Webster et al, 1997; p. 27)
    a. H1: Previous violence (0, 1, or 2 for None, Less, and More severe violence)
    b. H2: Young age at first violent incident (0 = 40+ years; 1=20-39; 2=<20)
    c. H3: Relationship instability (0=stable, conflict-free; 1=less serious unstable; 2=more serious unstable)
    d. H4: Employment problems
    e. H5: Substance use problems
    f. H6: Major mental illness
    g. H7: Psychopathy
    h. H8: Early maladjustment (home, school, community before age 17)
    i. H9: Personality disorder (esp. antisocial or borderline)
    j. H10: Prior supervision failure (e.g., parole, probation, other agency or institution)
2. **Clinical scale**: 5 Items related to current mental status.
    a. C1: Lack of insight ("*reasonable understanding and evaluation of one's own mental processes, reactions, self knowledge*")
    b. C2: Negative attitudes (e.g., towards others, social agencies, the law, other authorities; optimism/pessimism about the future; attitude towards past violent acts, etc.)
    c. C3: Active symptoms of major mental illness
    d. C4: Impulsivity ("*behavioral and affective instability*")
    e. C5: Unresponsive to treatment (getting help?; effort being put in, etc.)
3. **Risk management scale**: 5 items that assess future environmental factors.
    a. R1: Plans lack feasibility (i.e., staff plans for working with patient at and after discharge:

(1) delivery of more intense services to higher than to lower risk cases
(2) targeting of "*criminogenic needs*" (i.e., clinical risk factors)
(3) matching treatment modalities to offender's needs and learning styles
  b. R2: Exposure to destabilizers. (E.g., presence of weapons, substances, victim groups)
  c. R3: Lack of personal support. (E.g., patient, tolerant relatives and peers)
  d. R4: Noncompliance with remediation attempts. (E.g., medication and other therapeutic regimens)
  e. R5: Stress. (E.g., financial loss, deaths in family, etc.)

C. Final scoring: Recommended that no numerical cutoff be used, but scorer rate individual as at Low, Moderate, or High Risk based on clinical judgment. Some factors (e.g., psychopathy) may weigh higher in the clinical judgment than others.

D. Research on validity of HCR-20
  1. **Douglas et al (2003)**: Determined violence risk for 100 forensic psychiatric inpatients released into the community. Several clinicians rated each patient using HCR-20, and provided low-, medium- or high-risk judgments for each.
     a. Inter-rater reliability was .61 (kappa)
     b. AUC for total score related to physical violence was .70; for any violence was .67
     c. AUC was similar for final risk judgments, and adding final risk judgments to HCR-20 scores provided increased predictive validity
  2. **Douglas et al (2005)**: Compared HCR-20, VRAG, PCL-R, and Violent Offender Risk Assessment Scale (VORAS) on released Canadian prisoners randomly selected from those who recidivated violently, and those who did not.
     a. Found HCR-20 and VRAG useful in judging who would reoffend violently
     b. PCL-R and VORAS just OK (and significantly correlated)

VII. Sexual Violence Risk (SVR-20)

A. Similar to HCR-20 in construction, but designed for use with sex offenders

B. Authors (Boer et al, 1997) selected risk factors empirically related to future sexual violence.

C. Items chosen derived from review of literature on sex offenders

D. Little research since the tool is very new, though de Vogel et al (2004) in Dutch study found AUC for SVR-20 = .83

VIII. Level of Service Inventory-Revised (LSI-R): **Andrews & Bonta (2003)**

    A. 54 items grouped rationally (not empirically) into 10 scales. Includes both historical items and dynamic risk factors related to antisocial cognitions and social relationships.

    B. Subscales:
       1. Scale 1: Criminal history
       2. Scale 2: Education/employment
       3. Scale 3: Financial
       4. Scale 4: Family/Marital
       5. Scale 5: Accommodations
       6. Scale 6: Leisure/Recreation
       7. Scale 7: Companions
       8. Scale 8: Alcohol/Drug problems
       9. Scale 9: Emotional/Personal
       10. Scale 10: Attitudes/Orientation

    C. Originally developed to guide decisions about parole supervision for offenders serving sentences of less than 2 years, has demonstrated validity in predicting general and violent recidivism.
       1. **Bonta & Motiuk (1992):**
       2. **Kroner & Mills (2001):**
       3. **Gendreau, Goggin & Smith (2002):** Meta-analysis of PCL-R and LSI-R in predicting recidivism. Found LSI-R slightly better than PCL-R in predicting general and violent recidivism.

References

Andrews, D.A. & Bonta, J. *Level of Service Inventory-Revised (LSI-R)*. Toronto, ON: Multi-Health Systems, 2003.

Bonta, J. & Motiuk, LL. Inmate classification. *Journal of Criminal Justice*, 1992, 20, 343-353.

Cooke, D.J. & Michie, C. Refining the construct of psychopathy: Towards a hierarchical model. *Psychological Assessment*, 2001, 13, 171-188.

Douglas, K.S. , Ogloff, J.R.P, & Hart, S.D. evaluation of a model of violence risk assessment among forensic psychiatric patients. *Psychiatric Services*, 2003, 54(10), 1372-1379.

Douglas, K.S., Yeomans, M. & Boer, D.P. Comparative validity analysis of multiple measures of violence risk in a sample of criminal offenders. *Criminal Justice and Behavior*, 2005, 32(5), 479-510.

Gendreau, P., Goggin C. & Smith, P. Is the PCL-R really the 'unparalleled' measure of offender risk? A lesson in knowledge cumulation. *Criminal Justice and Behavior*, 2002, 29(4), 397-426.

Goldberg, L.R. Simple models of simple processes? Some research on clinical judgments. *American Psychologist*, 1968, 23, 483-496.

Harris, G.T. Rice, M.E. & Quinsey, V.L. Violent recidivism of mentally disordered offenders: The development of a statistical prediction instrument. *Criminal Justice and Behavior*, 1993, 20, 315-335.

Harris, G.T., Rice, M.E. & Quinsey, V.L. A multi-site comparison of actuarial risk instruments for sex offenders. *Psychological Assessment*, 2003, 15, 413-425.

Kroner, D.G. & Mills, J.F. The accuracy of five risk appraisal instruments in predicting institutional misconduct and new convictions. *Criminal Justice and Behavior*, 2001, 28(4), 471-489.

Monahan, J. *The clinical prediction of violent behavior*. Washington, D.C.: National Institute of Mental Health, 1981.

Monahan, J., Steadman, H.J., Clark Robbins, P., Appelbaum, P., Banks, S., Grisso, T., Heilbrun, K. Mulvey, E.P., Roth, L. & Silver, E. An Actuarial Model of Violence Risk Assessment for Persons With Mental Disorders. *Psychiatric Services*, 2005, 56, 810-815.

Monahan, J. & Steadman, H.J. Violence risk assessment: A quarter century of research. In Frost, L.E. & Bonnie, R.J. (Eds.) *The evolution of mental health law* (pp. 195-211). Washington, D.D.: American Psychological Association, 2001.

Preusse, M.G. & Quinsey, V.L. The dangerousness of patients released from maximum security: A replication. *Journal of Psychiatry and Law*, 1977, 5, 293-299.

Quinsey, V.L. & Ambtman, R. Variables affecting psychiatrists' and teachers' assessments of the dangerousness of mentally ill offenders. *Journal of Consulting and Clinical Psychology*, 1979, 47, 353-362.

Quinsey, V.L., Harris, G.T., Rice, M.E. & Cormier, C.A. *Violent offenders: Appraising and managing risk*. Washington, D.C: American Psychological Association, 2006. (Early version 1998)

Quinsey, V.L., Pruesse, M. & Fernley, R. Oak Ridge patients: Prerelease characteristics and postrelease adjustment. *Journal of Psychiatry and Law*, 1975, 3, 63-75.

Rice, M.E. & Harris, G.T. Cross-validation and extension of the violence risk appraisal guide for child molesters and rapists. *Law and Human Behavior*, 1997, 21, 231-241.

Salekin, R.T., Rogers, R. & Sewell, K.W. A review and meta-analysis of the psychopathy checklist and psychopathy checklist-revised: Predictive validity of dangerousness. *Clinical Psychology: Science and Practice*, 1996, 3, 203-213.

Steadman, H.J. & Cocozza, J. *Careers of the criminally insane*. Lexington, MA: Lexington Books, 1974.

Steadman, H.J., Silver, E., Monahan, J., Appelbaum, P.S., Clark Robbins, P., Mulvey, E.P., Grisso, T., Roth, L.H., & Banks, S. A classification tree approach to the development of actuarial violence risk assessment tools. *Law and Human Behavior*, 2000, 24, 83-100.

Thorberry, T. & Jacoby, J. *The criminally insane: A community follow-up of mentally ill offenders*. Chicago, IL: University of Chicago Press, 1979.

Psychopathy

I. Introduction

   A. Note that in our discussion of Violence Risk Assessment, Psychopathy - and the Psychopathy Checklist-Revised came up repeatedly:
      1. Included in the Assessment Results for the VRAG, and the best predictor of future violence in the final 12 variables at .34
      2. Included in the Iterative Classification Tree
      3. Included on the HCR-20

   B. Clearly one of the best predictors of future violent recidivism that we have.

   C. Because this is a forensic psychology course and not an abnormal psychology course, we will not be talking about etiology of psychopathy, only its nature, forensic detection, and treatment (a bit)

II. General Description and History

   A. Described frequently throughout history (certainly going back to the Greeks), and was mentioned by commentators and psychologists throughout the 18th and 19th centuries, including by Emil Kraepelin.
      1. Considered by many to be the founder of modern scientific psychiatry, psychopharmacology and psychiatric genetics; believed psychiatric diseases mainly caused by biological and genetic disorders.
      2. Kraepelin classified psychosis, into Manic Depression and Dementia praecox. (Schizophrenia)
      3. Colleague of Alois Alzheimer, and co-discoverer of Alzheimer's disease
      4. 1890s: Emil Kraepelin in the 2nd edition of his *Psychiatrie: Ein Lehrbuch* ("Psychiatry: A Handbook" talks about the "morally insane": Suffering from congenital defects in their ability to restrain "the reckless gratification of ... immediate egotistical desires." (Quoted in Millon et al, p. 9)

   B. Initial work on the concept done by Hervey Cleckley, M.D. in the 1930s and 1940s
      1. "The Mask of Sanity" published in 1942, and revised for a 1976 edition (Cleckley died in 1987). Most recent edition is 1988.

   C. Cleckley listed the 16 main characteristics of psychopathy:
      1. **Superficial charm and higher intelligence**: "the psychopath will seem particularly agreeable and make a distinctly positive impression when he is first encountered. Alert and friendly in his attitude, he is easy to talk with and seems to have a good many genuine interests. There is nothing at all odd or queer about him, and in every respect he

tends to embody the concept of a well-adjusted, happy person. ... He looks like the real thing. Very often indications of good sense and sound reasoning will emerge and one is likely to feel soon after meeting him that this normal and pleasant person is also one with high abilities. ... he is likely to seem free from social or emotional impediments, from the minor distortions, peculiarities, and awkwardnesses so common even among the successful." (Cleckley, p. 338)

2. **Absence of delusions and irrational thinking**: "He does not hear voices. Genuine delusions cannot be demonstrated. There is no valid depression, consistent pathologic elevation of mood, or irresistible pressure of activity. ... The results of direct psychiatric examination disclose nothing pathologic - nothing that would indicate incompetency or that would arouse suspicion that such a man could not lead a successful and happy life. Not only is the psychopath rational and his thinking free of delusions, but he also appears to react with normal emotions. ... He seems to respond with adequate feelings to another's interest in him and, as he discusses his wife, his children, or his parents, he is likely to be judged a man of warm human responses, capable of full devotion and loyalty." (Cleckley, p. 339)

3. **Absence of nervousness**: "the psychopath is nearly always free from minor reactions popularly regarded as "neurotic" or as constituting "nervousness." It is highly typical for him not only to escape the abnormal anxiety and tension fundamentally characteristic of this whole diagnostic group but also to show a relative immunity from such anxiety and worry as might be judged normal or appropriate in disturbing situations ... Even under concrete circumstances that would for the ordinary person cause embarrassment, confusion, acute insecurity, or visible agitation, his relative serenity is likely to be noteworthy." (Cleckley, p. 340)

4. **Unreliability**: "Though the psychopath is likely to give an early impression of being a thoroughly reliable person, it will soon be found that on many occasions he shows no sense of responsibility whatsoever. No matter how binding the obligation, how urgent the circumstances, or how important the matter, this holds true. ... If such failures occurred uniformly and immediately, others would soon learn not to rely upon psychopaths or to be surprised at their conduct. It is, however, characteristic for them during some periods to show up regularly at work, to meet their financial obligations, to ignore an opportunity to steal. ... Not all checks given by psychopaths bounce; not all promises are uniformly ignored. ... If so, it would be much simpler to deal with them. ... ... it cannot be predicted how long effective and socially acceptable conduct will prevail or precisely when (or in what manner) dishonest, outlandish, or disastrously irresponsible acts or failures to act will occur. ... Although it can be confidently predicted that his failures and disloyalties will continue, it is impossible to time them and to take satisfactory precautions against their effect. Here, it might be said, is not even a consistency in inconsistency but an inconsistency in inconsistency." (Cleckley, p. 340)

5. **Untruthfulness and insincerity**: "The psychopath shows a remarkable disregard for truth and is to be trusted no more in his accounts of the past than in his promises for the future or his statement of present intentions. ... Typically he is at ease and unpretentious in making a serious promise or in (falsely) exculpating himself from accusations, whether grave or trivial. ... Whether there is reasonable chance for him to get away with the fraud or whether certain and easily foreseen detection is at hand, he is apparently unperturbed and does the same impressive job. Candor and trustworthiness seem implicit

in him at such times. ... it is difficult to express how thoroughly straightforward some typical psychopaths can appear. They are disarming not only to those unfamiliar with such patients but often to people who know well from experience their convincing outer aspect of honesty." (Cleckley, p. 342-343)

6. **Lack of remorse or shame**: "The psychopath apparently cannot accept substantial blame for the various misfortunes which befall him and which he brings down upon others, usually he denies emphatically all responsibility and directly accuses others as responsible, but often he will go through an idle ritual of saying that much of his trouble is his own fault." (Cleckley, p. 343)

7. **Inadequately motivated antisocial behavior**: "Not only is the psychopath undependable, but also in more active ways he cheats, deserts, annoys, brawls, fails, and lies without any apparent compunction. He will commit theft, forgery, adultery, fraud, and other deeds for astonishingly small stakes and under much greater risks of being discovered than will the ordinary scoundrel. He will, in fact, commit such deeds in the absence of any apparent goal at all." (Cleckley, p. 343-345)

8. **Failure to learn from experience (poor judgment)**: "Despite his excellent rational powers, the psychopath continues to show the most execrable judgment about attaining what one might presume to be his ends. He throws away excellent opportunities to make money, to achieve a rapprochement with his wife, to be dismissed from the hospital, or to gain other ends that he has sometimes spent considerable effort toward gaining. ... Few more impressive examples of this could be offered from the records of humanity than the familiar one of the psychopath who, in full possession of his rational faculties, has gone through the almost indescribably distasteful confinement of many months with delusional and disturbed psychotic patients and, after fretting and counting the days until the time of his release, proceeds at once to get drunk and create disorder which he thoroughly understands will cause him to be returned without delay to the detested wards." (Cleckley, p. 345-346)

9. **Egocentricity and incapacity for love**: "The psychopath is always distinguished by egocentricity. ... probing will always reveal a self-centeredness that is apparently unmodifiable and all but complete. This can perhaps be best expressed by stating that it is an incapacity for object love and that this incapacity (in my experience with well-marked psychopaths) appears to be absolute. ... He is plainly capable of casual fondness, of likes and dislikes, and of reactions that, one might say, cause others to matter to him. These affective reactions are, however, always strictly limited in degree. ... psychopaths are sometimes skillful in pretending a love for women or simulating parental devotion to their children." (Cleckley, p. 346-348)

10. **General poverty in major affective reactions**: " In addition to his incapacity for object love, the psychopath always shows general poverty of affect. Although it is true that be sometimes becomes excited and shouts as if in rage or seems to exult in enthusiasm and again weeps in what appear to be bitter tears or speaks eloquent and mournful words about his misfortunes or his follies, the conviction dawns on those who observe him carefully that here we deal with a readiness of expression rather than a strength of feeling." (Cleckley, p. 348-350)

11. **Specific loss of insight**: "In a special sense the psychopath lacks insight to a degree seldom, if ever, found in any but the most seriously disturbed psychotic patients. ... His insight is of course not affected at all with the type of impairment seen in the

schizophrenic patient, who may not recognize the fact that others regard him as mentally ill but may insist that he is the Grand Lama and now in Tibet. Yet in a very important sense, in the sense of realistic evaluation, the psychopath lacks insight more consistently than some schizophrenic patients. He has absolutely no capacity to see himself as others see him." (Cleckley, p. 350-352)

12. **Unresponsiveness in interpersonal relationships**: "The psychopath cannot be depended upon to show the ordinary responsiveness to special consideration or kindness or trust. No matter how well he is treated, no matter how long-suffering his family, his friends, the police, hospital attendants, and others may be, he shows no consistent reaction of appreciation except superficial and transparent protestations." (Cleckley, p. 354-355)

13. **Fantastic and uninviting behavior with drink** (sometimes without): "Although some psychopaths do not drink at all and others drink rarely, considerable overindulgence in alcohol is very often prominent in the life story. ... A major point about the psychopath and his relation to alcohol can be found in the shocking, fantastic, uninviting, or relatively inexplicable behavior which emerges when he drinks - sometimes when he drinks only a little. It is very likely that the effects of alcohol facilitate such acts and other manifestations of the disorder." (Cleckley, p. 354-357)

14. **Threats of suicide rarely carried out**: "Despite the deep behavioral pattern of throwing away or destroying the opportunities of life that underlies the psychopath's superficial self-content, ease, charm, and often brilliance, we do not find him prone to take a final determining step of this sort in literal suicide. ... Instead of a predilection for ending their own lives, psychopaths, on the contrary, show much more evidence of a specific and characteristic immunity from such an act." (Cleckley, p. 358-359)

15. **Sex life is impersonal, trivial, and poorly integrated**: "In psychopaths ... there are varying degrees of susceptibility or inclination to immature or deviated sex practices. In contrast with others, the psychopath requires impulses of scarcely more than whim-like intensity to bring about unacceptable behavior in the sexual field or in any other. Even the faintest or most fleeting notion or inclination to forge a check, to steal his uncle's watch, to see if he can seduce his best friend's wife ... is by no means unlikely to emerge as the deed. The sort of repugnance or other inhibiting force that would prevent any or all such impulses from being followed (or perhaps from even becoming conscious impulses) in another person is not a factor that can be counted on to play much part in the psychopath's decisions." (Cleckley, p. 360-364)

16. **Failure to follow any life plan**: "The psychopath shows a striking inability to follow any sort of life plan consistently, whether it be one regarded as good or evil. He does not maintain an effort toward any far goal at all. This is entirely applicable to the full psychopath. On the contrary, he seems to go out of his way to make a failure of life." (Cleckley, p. 364)

D. "The [psychopath] is unfamiliar with the primary facts or data of what might be called personal values and is altogether incapable of understanding such matters. It is impossible for him to take even a slight interest in the tragedy or joy or the striving of humanity as presented in serious literature or art. He is also indifferent to all these matters in life itself. Beauty and ugliness, except in a very superficial sense, goodness, evil, love, horror, and humour have no actual meaning, no power to move him. He is, furthermore, lacking in the ability to see that others are moved. It is as though he were colour-blind, despite his sharp intelligence, to this aspect of human existence. It cannot be explained to him because there is nothing in his orbit of awareness that can bridge the gap with comparison. He can repeat the words and say glibly that he understands, and there is no way for him to realize that he does not understand" (Cleckley, 1941, p. 90 quoted in Hare, 1993, pp. 27-28).

III. Relationship between Cleckley's Psychopathy and the DSM Personality Disorders

   A. Does not appear as a diagnostic category in the DSM-IV, but closely related to Antisocial Personality Disorder (APD):
      1. "A pervasive pattern of disregard for and violation of the rights of others occurring since 15 years, as indicated by three (or more) of the following:
         a. "(1) failure to conform to social norms with respect to lawful behavior as indicated by repeatedly performing acts that are grounds for arrest"
         b. "(2) deceitfulness as indicated by repeated lying, use of aliases, or conning others for personal profit or pleasure
         c. "(3) Impulsivity or failure to plan ahead"
         d. "(4) irritability and aggressiveness, as indicated by repeated physical fights or assaults"
         e. "(5) reckless disregard for safety of self or others"
         f. "(6) consistent irresponsibility, as indicated by repeated failure to sustain consistent work behavior or honor financial obligations"
         g. "(7) lack of remorse, as indicated by being indifferent to or rationalizing having hurt, mistreated, or stolen from another"
      2. "The individual is at least age 18 years"
      3. "There is evidence of Conduct Disorder (see p. 98) with onset before 15 years."
      4. "The occurrence of antisocial behavior is not exclusively during the course of Schizophrenia or a Manic Episode."

   B. DSM-IV criterion for Antisocial Personality Disorder considered to be much broader than the criteria for psychopathy under PCL-R.
      1. Emphasize behaviors rather than underlying psychological processes.
      2. Minimal criteria for APD:
         a. "repeatedly performing acts that are grounds for arrest"
         b. "repeated lying, use of aliases"
         c. "repeated physical fights or assaults"

      3. Minimal criteria for APD:
         a. impulsivity, or failure to plan ahead
         b. irritability and aggressiveness
         c. reckless disregard for safety of self or others; or
         d. consistent irresponsibility, as indicated by repeated failure to sustain consistent work behavior or honor financial obligations

C. Psychopathy refers to a constellation of behaviors and measurable cognitive, emotional, and neuropsychological characteristics

IV. Statistics: How many people in the population would meed this criterion?

A. APD (ASP):
1. Refers to behavioral patterns based on clinical observation
2. Broad enough that 50-80% of Canadian prison inmates meet the criteria (**Correctional Services of Canada, 1990**; **Hare, 1998**); **Hare, Forth & Strachan, (1992)**
3. Only 11% - 25% of male inmates meet criteria for psychopathy (**Hare, 1996**; **Simourd & Hoge, 2000**)
   a. **Simourd & Hoge (2000)** found only 11% of 321 violent offender prisoners met the criteria.

B. Estimated 10-15% of child molesters are psychopaths; 40-50% among rapists (**Gretton et al, 2001; Porter et al, 2000**)
1. Michael Stone (in Millon et al, 1998): looked through 278 books on serial murderers and calculated that only 5% did now show all the symptoms of psychopathy - but sample was limited to spectacular cases written about.
2. **Hart & Dempster (1997)** suggest that psychopathic rapists more likely to have 'nonsexual' motivations for rape (anger, sadism, vindictiveness, etc.)

C. Hare estimates 1% of population fit criteria for psychopathy

V. Forensic issues with psychopathy

A. Murders and assaults by psychopaths less likely that those of nonpsychopaths to take place during domestic disputes or times of high emotional arousal. (**Hare et al, 1991**)

B. **Porter et al (2000)**: Psychopaths reoffend faster, violate parole sooner, and commit more institutional violence than nonpsychopaths.
1. **Serin & Amos (1995)** followed 299 male offenders for up to 8 years after release from federal prison. 65% of psychopaths convicted of another crime within 3 years, compared with 25% of nonpsychopaths
2. **Quinsey, Rice & Harris (1995)** found that within 6 years of release from prison, over 89% of sex offender psychopaths had violently recidivated; only 20% for nonpsychopathic sex offenders
3. **Serin, Peters & Barbaree (1990)** found that 37.5% of psychopaths failed on an unescorted temporary absence plan, while none of the nonpsychopaths failed.
   a. 33% of psychopaths violated parole requirements, only 7% of nonpsychopaths did

C. **Gretton et al (2001)**: Rates also high for psychopathic adolescent offenders who are more likely to:
1. Escape from custody
2. Violate conditions of probation
3. Reoffend (violently or not) over 5-year follow-up period

D. Individuals with diagnosis of psychopathy more likely to be found guilty - as is the case with diagnoses of APD/CD - more likely to be fund guilty in mock juror cases. **(Blais & Forth, 2014)**

VI. Psychopathy Checklist - Revised (PCL-R)

A. Assesses Robert Hare's conceptualization of psychopathy as a personality style that involves "*the remorseless use of others and subsequent irresponsible and antisocial behavior*" ((Jackson, p. 169)

B. Originally designed in the 1970s as a research tool for use with incarcerated and forensic cases. Although not designed for the purposes, has demonstrated utility in assessing future risk of violence

C. 20 items scored by clinician on 0, 1, 2 scale after file review, clinical interview, and sometimes collateral interviews:
1. Item #1: Glibness/superficial charm
2. Item #2: Grandiose sense of self-worth
3. Item #3: Need for stimulation/prone to boredom
4. Item #4: Pathological lying
5. Item #5: Conning/manipulative
6. Item #6: Lack of remorse or guilt
7. Item #7: Shallow affect
8. Item #8: Callous/lack of empathy
9. Item #9: Parasitic lifestyle
10. Item #10: Poor behavioral controls
11. Item #11: Promiscuous sexual behavior
12. Item #12: Early behavior problems
13. Item #13: Lack of realistic long-term goals
14. Item #14: Impulsivity
15. Item #15: Irresponsibility
16. Item #16: Failure to accept responsibility
17. Item #17: Many short-term marital relationships
18. Item #18: Juvenile delinquency
19. Item #19: Revocation of conditional release
20. Item #20: Criminal versatility

D. Scores of 30 or higher (sometimes 25 or higher) considered to indicate psychopathy

VII. Factor Analyses of the PCL-R

   A. Two Factor solution:
      1. Two higher-order factors:
         a. Interpersonal and affective traits
         b. Antisocial and impulsive lifestyle
      2. What's left out?
         a. Promiscuous sexual behavior
         b. Many short-term marital relationships
         c. Criminal versatility

   B. Three-factor solution:
      1. Three factors (**Cooke & Michie, 2001**)
         a. Interpersonal
         b. Affective
         c. Impulsive/irresponsible lifestyle
      2. What's left out?
         a. Promiscuous sexual behavior
         b. Many short-term marital relationships
         c. Criminal versatility
         d. Poor behavioral controls
         e. Revocation of conditional release
         f. Early behavior problems
         g. Juvenile delinquency

   C. Four-factor solution:
      1. **Interpersonal**
         a. glib, superficial
         b. grandiose self-worth
         c. pathological lying
         d. conning, manipulative
      2. **Affective**
         a. lack of remorse or guilt
         b. shallow affect
         c. callous/lacks empathy
         d. failure to accept responsibility
      3. **Lifestyle**
         a. stimulation seeking
         b. impulsivity
         c. irresponsibility

           d. parasitic orientation
           e. lack of realistic goals
        4. **Antisocial**
           a. poor behavior controls
           b. early behavior problems
           c. juvenile delinquency
           d. revocation of conditional release
           e. criminal versatility
        5. What's left out?
           a. Promiscuous sexual behavior
           b. Many short-term marital relationships

VIII. Validity of the PCL in its various versions

   A. Tons of studies, most support validity of PCL-R as predictor of violent recidivism

   B. Salekin, Rogers & Sewell (1996): Did meta-analysis of 18 studies of relationship between PCL-R and violence.
      1. Found that **behavioral characteristics** better predicted general recidivism
      2. Both **personality and behavioral aspects** related to violent recidivism
      3. Concluded "*Despite its limitations, the PCL-R appears to be unparalleled as a measure for making risk assessments with white males.*"

IX. Psychopathic Personality Inventory - Revised (PPI-R).

   A. The PCL (and later PCL-R) were designed as clinical assessment, primarily to be used on incarcerated populations, Scott Lilienfeld (1990) developed the Psychopathic Personality Inventory in 1990 for use with non-clinical, normal populations. It was originally developed on undergraduate student samples.

   B. The PCL-R was developed in 2005 (Lilienfeld & Widowes, 2005), and contains 154 items scored on eight subscales, based on three factors - two main factors, and a largely unrelated third factor:
      1. **Factor 1: Fearless Dominance**, includes the subscales for:
         a. **Social Potency** (interpersonal impact and skill at influencing others)
            (1) "Even when others are upset with me I can usually win them over with my charm."
         b. **Fearlessness** (a willingness to take physical risks and an absence of anticipatory anxiety)
            (1) "Making a parachute jump would really frighten me." - reverse coded
         c. **Stress Immunity** (sangfroid and absence of tension in anxiety-provoking

situations)
(1) "I can remain calm in situations that would make many other people panic."
2. **Factor 2: Impulsive Antisociality**, includes the subscales for:
a. **Machiavellian Egocentricity** (a ruthless willingness to manipulate and take advantage of others:
(1) "I sometimes try to get others o bend the rules for me if I can't change them any other way."
b. **Nonconformity** (a flagrant disregard for tradition)
(1) "I sometimes question authority figures just for the hell of it."
c. **Blame externalization** (a tendency to attribute responsibility for one's mistakes to others)
(1) "When I'm in a group of people who do something wrong, somehow it seems like I'm the one who ends up getting blamed."
d. **Carefree nonplanfulness** (an insouciant attitude towards the future)
(1) "I weight the pros and cons of major decisions carefully before making them." - reverse coded
3. **Factor 3: Coldheartedness**, includes only one subscale:
a. **Coldheartededness** (callouness, guiltlessness, and absence of empathy)
(1) "I have had 'crushes' on people that were so intense that they were painful." - reverse coded

C. While scores on the PPI-R and PCL-R are substantially correlated; **Copestake, Gray and Snowden (2011)** found correlations of .54 in a sample of male offenders, these authors also note that "the factors underpinning the PPI-R and the factors underpinning the PCL-R did not show any obvious correspondence", and suggest that the two tests measure different concepts of psychopathy and cannot be used as substitutes for one another. **Malterer et al (2010)** reached much the same conclusion about the relationship between the PCL-R and the PPI-R.

X. Treatment for psychopaths?

A. General feeling in the therapeutic community is that there is *"nothing the behavioral sciences can offer for treating those with psychopathy"* (**Gacono et al, 1997**, p. 119)
B. Psychopaths described as *"unmotivated to accept their problematic behavior and often lack insight into the nature and extent of their psychopathology"* (**Skeem, Edens & Colwell, 2003**, p. 26)
C. Some have gone so far as to argue that we should not treat psychopaths because it gives them skills to better deceive and manipulate others (mentioned on The Sopranos; try to find reference in medical journal)
1. **Hare (1996):**
2. **Rice, Harris & Cormier (1992)** examined records 10 years after participation in an intensive community therapy program in maximum-security facility, that psychopaths had higher rates of violent recidivism than psychopaths who did not participate. The

reverse was true for nonpsychopaths. Authors note that psychopathic sample was especially serious offenders, so results may not be typical.
    3. **Seto & Barbaree (1999)**
  D. But some others disagree (e.g., **Salekin, 2002; Skeem, Monahan & Mulvey, 2002; Skeem et al (in press); Wong, 2000**)
    1. **Skeem et al (2002)** found that psychopathic psychiatric patients who received 7 or more treatment sessions during a 10-week period were 3 times less likely to be violent than psychopaths who received fewer then 7 sessions.
    2. **Salekin (2002)** also found a range of interventions to be moderately success with psychopaths, especially if treatment was long and intensive.
  E. An even more recent paper by **Polaschek (2014)** reviews recent studies on therapeutic change in psychopaths and concludes that:
    1. "Psychopathic characteristics, especially those most related to criminal offending, can change over the life course
    2. "although adult criminals with psychopathy are among the hardest to work with in treatment, treatment causes them—like other offenders—to reoffend less"
    3. "There is no good evidence that criminals with psychopathy take advantage of treatment services to wreak havoc on therapists or the community.
    4. "Like other high-risk criminals, those with psychopathy can benefit from psychological treatment."

XI. Thoughts from Hans Toch (Distinguished Professor of Criminal Justice, School of Criminal Justice, SUNY Albany)

  A. Social psychologist in criminology and criminal justice administration. Wrote many books:
    1. "Living in Prison" ((Free Press, 1975)
    2. "Violent Men" (Aldine, 1969)
    3. "The Disturbed Violent Offender" (with Ken Adams, Yale, 1989)
    4. "Police as Problem Solvers" (with J.D. Grant, Plenum, 1991)
    5. "Mosaic of Despair" (A.P.A. Books, 1992)
    6. "Police Violence" (with William Geller, Yale, 1996)
    7. "Corrections: A Humanistic Approach" (Harrow and Heston, 1997)
    8. "Acting Out" (with Ken Adams), APA Books, 2002.

  B. Elected Fellow of American Psychological Association and American Society of Criminology; served in 1996 as President of American Association for Forensic Psychology. Consultant to the National Commission on the Causes and Prevention of Violence.

  C. "...the effects of (1) pathologizing chronic offense careers, (2) depathologizing other problems of the same offenders, (3) furthering (and later resuscitating) a link between violence and psychopathy, (4) advancing the notion of a criminal personality entity, and (5) postulating the non-amenability to intervention of chronic offenders, have been to provide scientific support for a psychologically dichotomous view of the human race ... and to promote a prescriptive inference that calls for the exiling or warehousing of serious delinquents and felons. .... I also believe that the focus on psychopathy has foreclosed questions that ought to be asked about the dynamics of offense behavior, especially in cases where offenses are serious, patterned, or repetitive. Such questions can only be asked if we center on the motives and perspectives of individual offenders, rather than assuming that criminals are essentially alike. Robert Hare's views ... notwithstanding, the phrase 'individual differences in personality' presupposes that there are differences among all offenders, which very much include differences among those who are now labeled psychopathic or antisocial." (Toch, in Millon et al, p. 145)

References

Blais, J. & Forth, A.E. (2014). Potential labeling effects: influence of psychopathy diagnosis, defendant age, and defendant gender on mock jurors' decisions. Psychology, Crime & Law, Vol. 20, No. 2, 116 134.

Cooke, D.J. & Michie, C. Refining the construct of psychopathy: Towards a hierarchical model. Psychological Assessment, 2001, 13, 171-188.

Copestake, S., Gray, N. S. & Snowden, R. J. (2011). "A comparison of a self-report measure of psychopathy with the psychopathy checklist-revised in a UK sample of offenders" (PDF). *Journal of Forensic Psychiatry & Psychology*, 22 (2): 169-182.

Correctional Services of Canada, 1990, *Forum on Corrections Research*, 2(1), Ottawa, ON: Author.

D'Silva, K., Duggan, C., & McCarthy, L. Does Treatment Really Make Psychopaths Worse? A Review of the Evidence. *Journal of Personality Disorders*, 2004, 18(2), 163-177.

Douglas et al (2005) - Reliability and Validity Evaluation of the Psychopathy Checklist: Screening Version (PCL:SV) in Swedish correctional and forensic psychiatric samples. *Assessment*, 2005, 12(2), 145-61.

Edens, J.F, Marcus, D.K., Lilienfeld, S.O. & Poythress, N.G. Psychopathic, not psychopath: Taxometric evidence for the dimensional structure of psychopathy. *Journal of Abnormal Psychology*, 2006, 115, 131-144.

Forth, A.E., Hart, S.D., & Hare, R.D. (1990). Assessment of psychopathy in male young offenders. *Journal of Consulting and Clinical Psychology*, 2, 342-344.

Forth, A.E., Kosson, D.S., & Hare, R.D. (2003). *The Psychopathy Checklist: Youth Version*. Toronto, Ontario: Multi-Health Systems.

Gacono, C.B., Nieberding, R.J., Owen, A., Rubel, J. & Bodholdt, R. Treating conduct disorder, antisocial, and psychopathic personalities. In J.B. Ashford, B.D. Sales, & W.H. Reid (Eds.) *Treating adult and juvenile offenders with special needs*. Washington, D.C.: American Psychological Association, 1997.

Gretton, H.M., McBride, M., Hare, R.D., O'Shaunessy, R. & Kumka, G. Psychopathy and recidivism in adolescent sex offenders. *Criminal Justice and Behavior*, 2001, 28, 427-449.

Guy Laura S. & Douglas, K. S. Examining the utility of the PCL:SV as a screening measure using competing factor models of psychopathy. *Psychological Assessment*, 2006, 18(2), 225-230.

Hare, R.D. Comparison of the procedures for the assessment of psychopathy. *Journal of Clinical and Consulting Psychology*, 1985, 53, 111-119

Hare, R.D. Psychopathy: A clinical construct whose time has come. *Criminal Justice and Behavior*, 1996, 23, 25-54.

Hare, R.D. Psychopathy, affect, and behavior. In D. Cooke, A. Forth & R. Hare (Eds.) *Psychopathy: Theory, research, and implications for society*. Dordecht, Netherlands: Kluwer. 1998.

Hare, R.D., Forth, A.E. & Strachan, K.E. Psychopathy and crime across the life span. In R.D. Peters, R.J. McMahon and V.L. Quinsey (Eds.) *Aggression and violence throughout the life span*. Newbury Park, CA: Sage, 1992.

Hare, R.D., Hart, S.D, & Harpur, T.J. Psychopathy and DSM-IV criteria for antisocial personality disorder. *Journal of Abnormal Psychology*, 1991, 100, 391-398.

Harpur, T.J, Hakstian, A.R. & Hare, R.D. Factor Structure of the Psychopathy Checklist. *Journal of Consulting and Clinical Psychology*, 1988, 56, 741-747.

Harpur, T.J., Hare, R.D. & Hakstian, A.R. Two-factor conceptualization of psychopathy: Construct validity and assessment implications. Psychological Assessment, 1989, 6-17.

Harris, G.T., Rice, M.E. & Quinsey, V.L. Psychopathy as a taxon: Evidence that psychopaths are a discrete class. *Journal of Consulting and Clinical Psychology*, 1994, 62, 387-397.

Hart, S.D. & Dempster, R.J. Impulsivity and psychopathy. In C.D. Webster and M.A. Jackson (Eds.) *Impulsivity: Theory, assessment, and treatment*. New York: Guilford, 1997.

Lewis, K., Olver, M.E. & Wong, S.C.P. (2013). The Violence Risk Scale: Predictive Validity and Linking Changes in Risk With Violent Recidivism in a Sample of High-Risk Offenders With Psychopathic Traits. *Assessment*, 20(2) 150–164.

Lilienfeld, S. O., & Widows, M. (2005). *Psychopathic Personality Inventory-Revised professional manual*. Odessa, FL: PAR.

Marcus, D.K. John, S.L., & Edens, J.F. A taxometric analysis of psychopathic personality. Journal of Abnormal Psychology, 2004, 113, 626-635.

Malterer, M.B., Lilienfeld, S.O., Neumann, C.S. & Newman, J.P. (2010) Concurrent Validity of the Psychopathic Personality Inventory With Offender and Community Samples. *Assessment*, 17(1) 3–15.

Polaschek, D.L.L (2014). Adult Criminals With Psychopathy: Common Beliefs About Treatability and Change Have Little Empirical Support. *Current Directions in Psychological Science*, Vol. 23(4) 296–301.

Porter, S., Fairweather, D., Drugges, J., Hervé, H., Birt, A. & Boer, D.P. Profiles of psychopathy in incarcerated sexual offenders. *Criminal Justice and Behavior*, 2000, 27, 216-233.

Quinsey, V.L., Rice, M.E. & Harris, G.T. Actuarial prediction of sexual recidivism. *Journal of Interpersonal Violence*, 1995, 10, 85-105.

Rice, M.E., Harris, G.T. & Cormier, C.A. An evaluation of a maximum security therapeutic community for psychopaths and other mentally disordered offenders. *Law and Human Behavior*, 1992, 16, 399-412.

Salekin, R.T., Rogers, R. & Sewell, K.W. A review and meta-analysis of the psychopathy checklist and psychopathy checklist-revised: Predictive validity of dangerousness. *Clinical Psychology: Science and Practice*, 1996, 3, 203-213.

Salekin, R. Psychopathy and therapeutic pessimism: Clinical lore or clinical reality? *Clinical Psychology Review*, 2002, 22, 79-112.

Serin, R.C. & Amos, N.L. The role of psychopathy in the assessment of dangerousness. *International Journal of Law and Psychiatry*, 1995, 18, 231-238.

Serin, R.C., Peters, R.D. & Barbaree, H.E. Predictors of psychopathy and release outcome in a criminal population. *Psychological Assessment*, 1990, 2, 419-422. .

Seto, M.C. & Barbaree, H.E. Psychopathy, Treatment Behavior, and Sex Offender Recidivism. *Journal of Interpersonal Violence*, 1999, 14(12), 1235-1248)

Simourd, D.J. & Hoge, R.D. Criminal psychopathy: A risk-and-need perspective. *Criminal Justice and Behavior*, 2000, 27, 256-272.

Skeem, J.L., Monahan, J. & Mulvey, E.P. Psychopathy, treatment involvement, and subsequent violence among civil psychiatric patients. *Law and Human Behavior*, 2002, 26, 577-603.

Skeem, J.L., Edens, J.F. & Colwell, L.H. *Are there racial differences in levels of psychopathy? A meta-analysis*. Paper presented at 3rd annual conference of the International Association of forensic Mental Health Services, Miami, Florida, 2003.

Skeem, J.L., Monahan, J. & Mulvey, E.P. Psychopathy, Treatment Involvement, and Subsequent Violence Among Civil Psychiatric Patients. Law and Human Behavior, 2002, 26(6), \

Skeem, J.L., Poythress, N., Edens, J., Lilienfeld, S. & Cale, E. Psychopathic personality or personalities? Exploring potential variants of psychopathy and their implications for risk assessment. *Aggression and Violent Behavior*, 2003, 8(5), 513-546.

Wong, S. Psychopathic offenders. In S. Hodgins & R. Muller-Isberner (Eds.) *Violence, crime, and mentally disordered offenders: Concepts and methods for effective treatment and prevention*. New York: Wiley, 2000.

Other Sources

Anestis, J.C., Caron, K.M. & Carbonell, J.L. (2011) Examining the Impact of Gender on the Factor Structure of the Psychopathic Personality Inventory–Revised. *Assessment*, 18(3) 340–349.

Drislane, L.E., Patrick, C.J. & Arsal, G. (2014) Clarifying the Content Coverage of Differing Psychopathy Inventories Through Reference to the Triarchic Psychopathy Measure. *Psychological Assessment*, Vol. 26 (2), 350–362.

Edens, J.F. & McDermott, B.E. (2010) Examining the Construct Validity of the Psychopathic Personality Inventory–Revised: Preferential Correlates of Fearless Dominance and Self-Centered Impulsivity. *Psychological Assessment*, 22 (1), 32–42.

Gonsalves, V.M., McLawsen, J.E., Huss, M.T. & Scalora, M.J. (2013) Factor structure and construct validity of the psychopathic personality inventory in a forensic sample. *International Journal of Law and Psychiatry*, 36, 176–184.

Hughes, M.A., Stout, J.C. & Dolan, M.C. (2013). Concurrent Validity of The Psychopathic Personality Inventory–Revised and The Psychopathy Checklist - Screening Version in an Australian Offender Sample. *Criminal Justice and Behavior*, Vol. 40 (7), 802-813.

Morgan, J.E., Gray, N.S. & Snowden, R.J. (2011) The relationship between psychopathy and impulsivity: A multi-impulsivity measurement approach. *Personality and Individual Differences*, 51, 429–434.

Patrick, C. J., Fowles, D. C., & Krueger, R. F. (2009). Triarchic conceptualization of psychopathy: Developmental origins of disinhibition, boldness, and meanness. *Development and Psychopathology*, 21, 913–938.

Polaschek, D.L.L. & Daly, T.E. (2013) Treatment and psychopathy in forensic settings. *Aggression and Violent Behavior*, 18, 592–603.

Scott, R. (2014) Psychopathy - An Evolving and Controversial Construct. *Psychiatry, Psychology and Law*, Vol. 21 (5), 687-715.

Sellbom, M., Wygant, D.B. & Drislane, L.E. (2015) Elucidating the Construct Validity of the Psychopathic Personality Inventory Triarchic Scales. *Journal of Personality Assessment*, 97(4), 374–381.

Toney Smith, S., Edens, J.F. & McDermott, B.E. (2013) Fearless Dominance and Self-Centered Impulsivity Interact to Predict Predatory Aggression among Forensic Psychiatric Inpatients. *International Journal of Forensic Mental Health*, 12: 33–41.

Uzieblo, K., Verschuere, B., Van den Bussche, e. & Crombez, G. (2010). The Validity of the Psychopathic Personality Inventory–Revised in a Community Sample. *Assessment*, 17(3) 334–346.